Quarterly Essay

CONTENTS

Quarterly Essay is published four times a year by Black Inc., an imprint of Schwartz Publishing Pty Ltd
Publisher: Morry Schwartz

ISBN 1 86395 227 6

Subscriptions (4 issues): $46.95 a year within Australia incl. GST (Institutional subs. $52.95). Outside Australia $74.95. Payment may be made by Mastercard, Visa or Bankcard, or by cheque made out to Schwartz Publishing. Payment includes postage and handling.

To subscribe, fill out and post the subscription form on the last page of this essay, or subscribe online at:

www.quarterlyessay.com

Correspondence and subscriptions should be addressed to the Editor at:
Black Inc.
Level 5, 289 Flinders Lane
Melbourne VIC 3000 Australia
Phone: 61 3 9654 2000
Fax: 61 3 9654 2290
Email: quarterlyessay@blackincbooks.com
http://www.blackincbooks.com

Editor: Peter Craven
Management: Silvia Kwon
Managing Editor: Chris Feik
Production Coordinator: Sophy Williams
Publicity: Meredith Kelly
Design: Guy Mirabella
Printer: Griffin Press

Quarterly Essay aims to present significant contributions to political, intellectual and cultural debate. It is a magazine in extended pamphlet form and by publishing in each issue a single writer at a length of at least 20,000 words we hope to mediate between the limitations of the newspaper column, where there is the danger that evidence and argument can be swallowed up by the form, and the kind of full-length study of a subject where the only readership is a necessarily specialised one. *Quarterly Essay* aims for the attention of the committed general reader. Although it is a periodical which wants subscribers, each number of the journal is the length of a short book because we want our writers to have the opportunity to speak to the broadest possible audience without condescension or populist short-cuts. *Quarterly Essay* wants to get away from the tyranny that space limits impose in contemporary journalism and we give our essayists the space to express the evidence for their views and those who disagree with them the chance to reply at whatever length is necessary. *Quarterly Essay* will not be confined to politics but is centrally concerned with it. We are not interested in occupying any particular point on the political map and we hope to bring our readership the widest range of political and cultural opinion which is compatible with truth-telling, style and command of the essay form.

A few weeks ago when the Greens won the federal seat of Cunningham what had been obvious to many became evident to all: the Greens were not a marginal fly-by-night party, they were a significant force in Australian politics. In this new and timely *Quarterly Essay* Amanda Lohrey attempts to anatomise the Greens, to show where they have come from and where they are going.

In the process she makes it clear that the Greens are a political party (and beyond that a political formation) that is built for long continuance and in this respect she sees them as in marked contrast to the Democrats with their fortuitous oppositionalism as a breakaway grouping. In Lohrey's analysis the Greens not only come out of the Western self-scrutiny that was part of the counterculture thirty years ago, the green movement represents the most coherent and considered product of that period and one which has been tested and proved true in the subsequent decades.

She speaks as a Tasmanian steeped in Labor Party politics and in the peculiar resonance that environmentalism took on in the seventies in Tasmania where the very concept of 'wilderness' became a talisman of what it meant to take the progressive side in Australian politics.

At the same time, Amanda Lohrey is a political thinker with the sweeping eye of a novelist and she is alive to the complexities of the past and its unwillingness to conform meekly to our retrospective pieties. Her hymn to the honour of the Greens is all the more impressive for the fact that she understands what it was like to be brought up in the fifties and sixties, on the myth of a progressive science that could transform the world, as surely for God's little Australians as for Soviets who saw the waters of their dams as sacramental. And she acknowledges too the tough old Labor men to whom the fight to save the Franklin was a sentimental folly, a conundrum that made no sense in a world of jobs and struggle and getting by.

Amanda Lohrey is one of those rare writers who can get inside the pastness of the past in order to illuminate the present. She talks (with enthusiasm) of the early green protesters setting themselves to learn the classic techniques of Gandhian civil disobedience and she also delineates the way some of the faithful recoiled from the 'touchy-feely' emphasis of the times.

She is equally good in establishing the way the green movement enters the mainstream of Australian politics and there is something like a tribute to that most improbable hero of environmentalism, Graham Richardson, who proved to be a dynamic force of nature when he was Minister for the Environment and saved Bob Hawke's bacon in the process.

Richo is a crucial figure for Amanda Lohrey because he was an undisputed master of *vox populi* and he was not wrong to see the depth of the Australian people's concern for the environment.

Lohrey is casually scathing about the track record of Keating's government but she has other fish to fry. Her central contention is that the Greens are people with a history, an archaeology and a lucid political faith that tends to give them the strength of religious illumination. They are people who have transcended the romanticism of tree hugging and wilderness mysticism (though their spirituality about the natural world is part of their tacit strength) in order to embrace a coherent political philosophy which begins with a scientific critique of the excesses of capitalism but has developed into the major force in this country, on the left, that sets itself against the go-for-broke world of market economics with its disdain for human consequences or consequences for the future of the Earth. In that respect Amanda Lohrey sees the Greens as the left-wing complement of Pauline

Hanson's One Nation, since both movements stand in defiance of the middle of the road marketeering that links the Labor and Coalition in an embrace that makes clear that they have both consented to a world view dominated by economic rationalism and are happy to live with the consequences.

One consequence, made clear at the level of liberal feeling in the wake of the *Tampa* episode and the treatment of the refugees, is that the Labor Party is being actively apprehended by its old adherents as what it has been for the better part of two decades, a party of the Right.

Amanda Lohrey sees this with great clarity but she sees the coherence of the Green alternative and the penetration of its grass-roots support as a far deeper thing than the protest vote, born of disgust and compassion, at the last election. Lohrey has been watching the figures and she has a sustained, impressive analysis of election results and she does her best to demonstrate (in a way that will appal Labor and Liberals alike, let alone the Democrats) that the increase in the Green Vote is exponential and that they are set to be the major player among the minor parties. In short that their visibility at the last election was not the flash in the pan that some people thought it, that they command a latent depth of respect across the community and that the broad affiliates of their supporters are, in fact, much greater numerically than the active memberships of the major political parties.

Lohrey writes about this with a controlled intellectual excitement that is everywhere alive to the realities of the revolution that has brought a group of people from rural and often far from modish backgrounds into alliance together. She is canny about the Greens as the reverse mirror image of One Nation and she is sufficiently inward with the players to cotton on to the significance of the fact that the Australian Greens are often people from the Bush with a fair measure of Christian conviction somewhere in the background. She realises the complexity of what goes to make up a political faith and she comprehends the fact that if the Greens are a party that command zeal and speak with moral authority then there are some old winds of the spirit blowing in the back of them and this in turn makes explicable the recruits they can make among people who are in significant ways both Christian and conservative.

Bob Brown in Amanda Lohrey's characterisation is certainly a man for all seasons. She emphasises the scepticism as well as the spirituality and the kind of personal integrity that can hush a House of Parliament by force not of charisma but of conviction. It is a portrait by a woman who is everywhere sensitive to the complexity and collective character of all political action and achievement, a novelist who is still discernibly a woman of the Left (and in that sense an Old Believer), but it is a testament to one man's personal distinctiveness and aura of honour. And, as it happens, it tallies with the approbation of ordinary Australians, Labor and Liberal, during the dismal lead-up to the last election, when there was the widespread perception, among the dissenters, that this gawky bespectacled gay doctor from Tasmania was some kind of hero. Older Australians will be aware of what it means to say that someone is 'close to God'. Well, Bob Brown, as Amanda Lohrey presents him, is close to *something*.

She also sees the Australian political movement of which he is the leader as close to the heart of the mystery of what ethical political action can be. She sees the green movement as a political philosophy rational to its core (but imbued with the conviction of religious reformism) that offers a total view of the physical universe in ways that encompass a post-humanist perspective because they envision the long-term good of the earth and not just the interests of the human beings who happen to populate it at this instant. Because this is a perspective grounded in a predictive scientific sense of the future it is also a way of cutting through the kind of postmodern binds that say all truths are relativistic, constructed things.

And because the Greens, both as an international movement and in their Australian formation, are not just liberals and humanists, this — paradoxically — gives a greater rigour and authority to the moral analysis and political scepticism they bring to bear on the opportunistic consensualism of the major parties.

This is an impassioned *Quarterly Essay* which is also a groundbreaking account of why the Greens are here to stay and why they are looking like a good thing. Amanda Lohrey understands the faith that animates the Greens and the history that sustains them and she also has the coolness of mind and the analytical equipment to show just how they lie on a political map that has changed beyond recognition in the last twenty years. This is the first essay we have published by someone who is primarily a writer of fiction: I hope it says something for the *Quarterly Essay* endeavour that Amanda Lohrey's essay is in the best traditions of the political essay and concedes nothing in the way of rigour or factual command to any of its predecessors.

Peter Craven

'Now and then,' wrote the great American critic Lionel Trilling, 'it is possible to observe the moral life in process of revising itself, perhaps by reducing the emphasis it formerly placed upon one or another of its elements, perhaps by inventing and adding to itself a new element, some mode of conduct or feeling which hitherto it had not regarded as essential to virtue.' Trilling's words come from a series of lectures he gave in 1970 entitled *Sincerity and Authenticity* and it's to this notion of authenticity and the moral life revising itself that the Green movement speaks today. It's not that authenticity (a congruence between avowal and feeling, or as we say now, an ability to walk the talk) is newly valued in our culture. It has for a long time been regarded as 'essential to virtue' but it is only in recent times that the feeling has arisen in the community of an identikit politics acted out by identikit politicians who bear little or no organic relation to the grass roots they claim to represent – on either the Left or the Right – and who have lost their moral compass when confronted by the challenges of globalisation. The rise of One Nation has been one response to this phenomenon: the rise of the Greens is another.

The trajectory of the Australian Greens Party in national politics is a big subject and I cannot hope to do justice to it here. It is, however, a long-standing interest of mine and I have given an unashamedly personal account of it, though one borne out by analysis. As a Tasmanian I have observed at close quarters some crucial moments in its evolution. Though a small state in both size and population, when it comes to environmental politics Tasmania has long been the canary in the mine, and in any case it's an old political adage that the earliest stirrings of profound change more often come from the margins than from some putative centre.

At the time of conceiving this essay I was responding to the Green vote in the 2001 federal election which was widely interpreted as a soft vote, a situational protest vote that would soon bounce back. It seemed to me, however, to be a portent of something more and indeed, in the time it has taken me to write this piece the electoral gains of the Greens have kept ahead of me. Just a week before this essay went to press they won their first ever federal seat in the House of Representatives, humiliating the ALP in the hitherto safe industrial seat of Cunningham. As a result of this unprecedented victory it is reasonable to assume that more people than ever will be asking the question: who are the Australian Greens and where do they come from? I hope this essay will go some way towards enlightening them.

Amanda Lohrey

Amanda Lohrey

In September of 1998 I sat in the Stanley Burbury Theatre at the University of Tasmania listening to a light-hearted debate in which Bob Brown told the story of how he had first arrived in Tasmania. It was 1972 and Brown was twenty-seven, a young doctor who had come from Sydney via New Zealand to work for six weeks as a locum in a busy practice in the northern Tasmanian city of Launceston. Called out late one night to an outlying town that he scarcely knew, he drove around the unfamiliar streets for some time trying without success to locate his 'call'. Finally he decided to ask for directions and stopped outside one of the few houses with its lights still on. When he approached the front door and knocked, a weak voice from inside called, 'Come in.' Brown opened the unlocked door and found a man writhing in pain on the floor.

'I'm Bob Brown,' he said. 'I'm a doctor.'

'Well,' gasped the man on the floor, 'you're the best bloody doctor I've struck. I called two doctors and none of the bastards came and I didn't call you and you've come anyway.'

Since then this story of the doctor who wasn't called but came anyway has stayed with me as somehow quirkily apt in its capture of the evolving dynamic that we now refer to as the green movement. It's also the perfect introduction to the enigmatic man who has become its national leader.

But if Bob Brown and his cohorts are the 'doctor', then what exactly is the 'ailment' and wherein lies the 'cure'?

Recent elections suggest that the Australian Greens Party is about to supersede the Democrats as a third political force in Australian politics. The reasons for this are both various and compelling. Though the fortunes of the Greens have fluctuated in federal elections over the past decade, it nevertheless remains the case that taken overall their support has grown steadily. Unlike both the Democrats and One Nation, the Greens are an organic party in the sense that they have evolved over a lengthy period of time and out of several community campaigns organised at the grass roots. Although for some years they have had strong personal leadership in the form of Bob Brown, unlike the Democrats and One Nation the Greens Party was not founded through the personal fiat of one individual. It has a strong history and folklore that pre-dates its leader and was a forceful influence in shaping him and the character of his leadership. Though Brown is not to be underestimated as a factor in the success of the Greens Party, even more telling is the broadening out of its base and the fact that over the past two decades it has evolved into a real constituency, something more than just a broad-based protest vote.

This constituency has been shaped by certain social and political changes of the last thirty years that have led to the emergence of a new political sensibility. Within this sensibility is a spectrum that ranges from right-wing tokenism to radical pantheism, but the extent of it is reflected in the no doubt perfectly sincere if somewhat risible remark of former federal Liberal Minister for the Environment Robert Hill to the effect that

'Everyone now is an environmentalist.' Corporations are for the most part routinely called upon to present Environmental Impact Statements, however fitted up, and no local council debate is complete without some genuflection towards the clean and the green.

Whereas the Democrats have always been a broad-based protest party, the new Green constituency is based not just on individual policy items, such as the preservation of old growth forests, but on a new paradigm or grand narrative of what politics is about, i.e. the 'ecological'. This of course is not a local phenomenon and has become the basis of a proliferating number of Green parties across the world which collectively amount to a significant component of the global Green alliance. This movement and its ecological narrative have the power to subsume the traditional grand narratives of capital and labour and indeed to some degree already have. It is not a force with which, in the long term, the Democrats can hope to compete. The Democrats have never had a natural constituency as such, some core element for whom they represent a set of primary values. They did not evolve from a grass-roots movement. In emerging from the moderate Right they offered a safe protest vote for disenchanted Liberal voters and in their stress on participatory democracy and critique of elite castes and party machines they also appealed to progressive or Left elements in the community. But always within the Democrat vote there was an early green vote that had nowhere else to go. Indeed the first green activist elected to an Australian parliament, Dr Norm Sanders (Tasmanian House of Assembly, 1980), joined the Democrats – and subsequently served as a Democrat senator – because there was at that time no Greens party to join. Sanders was and is a flamboyant and outspoken character whose most famous pronouncement in relation to his new party was to undermine its slogan by saying, 'If we get into power we'll be just as big a pack of bastards as the rest of them.' Recent events appear to confirm the acuity of his remark, although it is not my intention here to take a cheap shot at the Democrats. Their record on the environment, particularly their early record, has been for the most part honourable. Nevertheless, in federal elections over the past decade their vote has been declining, even before the unhappiness occasioned by their support for the GST. Single issues like the GST come and go on the political horizon but what is more important over time are changes at the deeper level, and the new Green constituency grows out of just such a deep paradigm shift in what the postmodernists like to call the social imaginary, i.e. the parameters of what can be conceived of in a society and the evolving myths it has of itself. It bears an important relation to what the British social theorist Raymond Williams once referred to as the dominant 'structures of feeling' that permeate a culture in any given era. By this Williams meant the area of tension between ideology and primary experience; between the official consciousness of an epoch, as codified in its doctrines and legislation, and the whole process of actually living out its consequences. It's in this gap that new political sensibilities begin to evolve.

Since the time of the industrial revolution a key element in the official ideology of Western culture had been the utopian potential of science but much of the lived experience of individuals had involved an increasing dislocation of the self from nature. One of the earliest outcomes of that sense of physical and psychological estrangement was the Romantic movement. The Romantics looked back to the past and to some idealised pre-industrial Eden, but the manifest ability of science to improve the individual's quality of life, not least through modern medicine, eventually put paid to the Romantic critique of Progress. It was not until the sixties that signs began to emerge of the development of a new 'ecological sensibility' and of a different kind of critique of industrial civilisation from that of the Romantics, one that arose out of a radical re-visioning of the future – not as a utopia produced by technology but as a potential wasteland of ecological disaster.

From the sixties on a number of eco-philosophers began to identify and articulate what Peter Hay has described in *Main Currents in Western Environmental Thought* as an 'ecological impulse', which he defines, in part, as an 'instinctive dismay' at the effects of technological advances upon nature. 'The wellsprings of a green commitment', writes Hay, 'are not, in the first instance, theoretical; nor even intellectual. They are, rather, pre-rational.' This pre-rational impulse is 'a deep-felt consternation at the scale of the destruction wrought, in the second half of the twentieth century, and in the name of transcendent human progression, upon the increasingly embattled life-forms with which we share the planet. It is an instinctive and deep-felt horror.'

The term 'ecological sensibility' is first deployed in the early seventies by the American political philosopher John Rodman. Defining sensibility as 'a complex pattern of perceptions, attitudes and judgements' that con-stitute 'a *disposition to appropriate conduct* that would make talk of rights and duties unnecessary' (my italics), Rodman lists among the attributes of this sensibility 'a style of cohabitation that involves the knowledgeable, respect-ful, and restrained use of nature'. For several decades there had been a reverence for wilderness in North America promoted by the great natural-ists like John Muir but this, like the Romantic movement, placed a strong emphasis on the past and the preservation of some idealised primal land-scape. It wasn't until the sixties that a new and radical critique emerged from within science itself to identify and oppose the destructive effects of pollution on a viable future.

That critique can be dated from the publication of one of the most sem-inal books of the twentieth century, Rachel Carson's *Silent Spring* (1962). Carson was a zoologist, and her observation of the effects of chemical pol-lutants on animal and bird life produced the first wave of recoil — the 'instinctive ecological compassion' that Hay refers to — that was to kick-start the modern environmental movement and its key ecological insight of the interconnectedness of all life. In the decades that followed, Man-within-Nature came to be seen not in the pastoral mode — nature as benign backdrop that can mostly be taken for granted — but as a more dynamic model of interconnectedness and interdependence. It's around this time that notions of sustainable development emerge, and economic, political and social theory begins to swerve away from a central concern with human relations and towards the physical world as the necessary basis for social and economic policy.

The political potency of ecology – its potential to shape a mainstream constituency – derives from this key ecological insight about the interconnectedness of all life, namely that we exist in and are dependent on a system of relationships that are universal. This interconnectedness is something that is increasingly demonstrated by scientific findings about, say, global warming, and as science mounts its own critique of industrialism it offers a more and more coherent and convincing articulation of the 'ecological impulse'; that latent disposition of the human mind towards preservation of the environment which eco-philosophers have argued for as a defining quality of human nature. What has hitherto been experienced as an inchoate sense of connectedness emerges – through scientific articulation – as a basis for programmatic action, since with an increase in intellectual understanding comes an enhanced sense of personal accountability (about even such 'simple' everyday things as garbage disposal and taking the time to sort recyclable items).

The deepening realisation that everything in the material universe is connected with everything else has produced at the political level a kind of green prism through which, increasingly, every aspect of politics is filtered. The more connections are made about the impact that one area has on another thousands of miles away – or the consequences one action will have on the parameters of action a long time in the future – the more the ecological constituency grows. This is happening in Australia as awareness focuses on an ever-expanding set of crises about such things as salinity, water conservation and blue-green algae, not to mention the effects of global warming. Its impact can be felt within the building and energy industries where greater emphasis is now being placed on energy conservation and the wind generation of electricity. And it is happening in what is being hailed as the next industrial revolution – ecological or sustainable capitalism.

The ethical responses to this are diverse. The deep ecologists, for example, despise the varieties of instrumental environmentalism that place immediate human welfare firmly at the apex of a hierarchy of needs. Philosophical differences aside, however, it is clear that at the level of everyday community concern there is a growing political constituency of great heterogeneity, made up of individuals and groups that cut across traditional class, economic and regional divides. In Australia this new Green constituency includes both disenchanted small 'l' Liberals and radical social justice reformers; investment analysts and kindergarten workers; accountants and ferals; Christians and Buddhists. There are revisionist farmers who see a new market in Europe for 'clean' produce; workers in the corporate sector who have initiated green investment strategies and trade unionists who have begun to understand that smog is not democratically distributed and the dumping of workers onto polluted housing sites is as big a problem as workplace health and safety. There are doctors and nurses who see an exponential growth in health problems arising out of pollution and teachers who have to keep a special drawer in their desks for the asthma puffers of so many of their small charges. These people have added their voices to a protest movement that began with water and forests and has grown into a widening critique of corporate values in the era of globalisation.

The political analyst and environment historian Tim Doyle has set forth a palimpsest model of how this constituency operates at the level of grassroots activism, made up at any one time of hundreds of local organisations claiming to espouse environmental values. Doyle defines the palimpsest as 'a series of amorphous networks with no overriding collection of goals … a strong potential constituency at any moment that does not depend on a base'. It includes both formal organisations, informal groups and individuals and 'is a snap-shot in time of a social movement that is alive, always moving, always redefining its shape and force … Although the movement does not have a common goal, each network does. The intermeshing of these networks makes up the movement.' It may involve Doctors for

Forests, your local Landcare group or Clean Up Australia committee, the WWF Great Barrier Reef campaign or any one of a constellation drawn from, on Doyle's estimate, a staggering 20,000-plus groups with a total active membership well in excess of the combined membership of all political parties in Australia.

As the palimpsest model demonstrates, a constituency is a social context that exists naturally; it arises and evolves without being manufactured — that is its strength — whereas it is in the nature of representation (parliamentary representatives, labels, symbolic and organisational structures) that it comes after and is secondary to the emergence of what has hitherto been latent. This is the political potency of the Australian Greens Party, that it has arisen out of and continues to evolve from its grass roots — from concerns that developed out of the cultural revolution of the sixties and seventies, and the new sensibility I have alluded to — and that it has grown in strength and scope from one campaign to the next.

In the history of its emergence, certain of these campaigns stand out as seminal.

The story of Lake Pedder has been told before and is well known to anyone over forty. Nevertheless it is worth revisiting here, not least because if there is a foundational narrative of the Greens Party in Australia it is the flooding of Lake Pedder by the Tasmanian Hydro-Electric Commission in 1972. The story of Lake Pedder and its erasure are to the Greens what the great shearers' strike and the town of Barcaldine in the 1890s were to the Australian Labor Party. This is the Greens' own Genesis story and Pedder is its paradise lost.

It's one of the more suggestive ironies of the Australian landscape and the political imagining of that landscape that of the two lakes which are best known internationally, one, Lake Eyre, is empty while the other, Lake Pedder, is submerged. When the early European explorers trekked into the centre looking for the inland sea that would fertilise a continent they found instead a vast empty lake-bed with a glittering white crust of salt that blinded them against the horizon. The Arabana people called it Katitanda but it soon became known as Lake Eyre and for a long time its enduring image was one of desolation and death. In contrast, the effect of Lake Pedder on its early visitors was to induce a kind of euphoric serenity. Unlike Eyre, this remote alpine lake in the heart of the south-western Tasmanian wilderness was a paradisiacal site of plenitude that came to have a mystical significance for those who beheld it. 'No description', wrote the painter, Max Angus, 'could convey the sense of awe and wonder felt by all those who saw this magic place.'

Unique in its formation and its beauties, Lake Pedder was nine square kilometres of water nestling between the quartzite peaks of the Coronets and the Frankland Range. Famed for the amber colour of its water but even more for the dazzling expanse of pale pink-white quartzite sands, the lake was proclaimed a national park in 1955. It was a unique ecosystem with several plants and more than a dozen creatures found nowhere else on earth – caddis flies, weird fossil-like shrimps, miniature snails, aquatic worms, a freshwater crayfish and *Galaxias pedderensis* – the Pedder trout – as well as wombats, Tasmanian devils, wallabies, native quolls, platypuses, marsupial mice, possums, tiger cats, black swans, rare ground parrots and emu-wrens.

In 1972 all of this was submerged beneath a shroud of water. Australia lost a whole ecosystem that it knew almost nothing about.

At the time I was an active member of Labor Youth and I supported the flooding of the lake, as did most of my political cohort. Though we were in the early stages of a cultural shift that was significantly to change us, at that time we still believed in the 'drive towards modernisation', as it was then called. The word 'modern' has become dated now, almost quaint, but back then it had a talismanic quality; the magical properties of the straight line and the sharp edge.

No one who wasn't there in the fifties or sixties can imagine the hegemony of science that prevailed in the post-war era. Science was the official state religion. Younger generations may gape now in amazement at Edward Teller's proposal to blast a harbour out of the Alaskan coast using a nuclear bomb but then it seemed merely radical rather than seriously mad. We, the baby-boomers, were a generation raised on the idea of science as intellectual and moral heroism, a status now accorded only to medical researchers (and with the advent of bio-technology, perhaps not even to them). As a child I can remember being taken on a class tour of the Tasmanian hydro-electric power stations – Waddamana, Tarraleah, Wayatinah – to gaze with appropriate reverence at their massive cables and turbines (the dams themselves were miles inland in the heart of the wilderness). When, many years later, I watched a documentary on the USSR, I had no trouble identifying with a particular sequence set in one of the Soviet provinces strong in hydro power. This showed how it was the practice of married couples to drive out to the local dam after the wedding ceremony in order to toss the bridal bouquet into the waters over the spillway. It hadn't quite come to that in Waddamana, or in the Snowy Mountains for that matter, but in either case the mythologising had been almost as grandiose and the politicians just as zealous. Technology engendered its own high romance, its own charismatic priesthood.

It was in the spirit of this ethos that throughout the fifties Tasmanian governments fostered a strong hydro-electric culture led by Sir Allan Knight, an engineer of genius. Knight and his team were in the vanguard of a project at once romantic and utilitarian. The dams were there to produce power in excess of the needs of the population so that it could be sold off cheaply to large industry (sometimes secretly at prices below cost) as a supposed guarantee of 'jobs'. At the same time the dams were an awe-inspiring marvel, something to be proud of, and the Tasmanian Labor politicians of the day *were* proud of them, seeing in the dams an emblem of how radical public spending could help lift a community out of the misery of the Depression years and into a better world. Many of these men were organic Labor politicians in the mould of the late Jack Ferguson, self-educated men who had come up from unskilled, working-class backgrounds of real poverty and they were genuinely baffled by the clamour over the flooding of Lake Pedder. I can recall having a conversation with a state cabinet minister, a man I greatly admired then, and still do. He asked me whether I could explain to him 'what made these people [the Pedder activists] tick'. 'The Hydro engineers are all nature lovers,' he told me. 'Many of them go bushwalking every weekend.' In addition, he explained, whereas before the dam, the lake had been remote and difficult to access, now the government had created a first-class road so that not just an elite few but ordinary people in their thousands could drive to the spot and enjoy the spectacular scenery.

'Scenery' was a word in more common usage then than it is now and the earliest protection agencies were entitled Scenery Preservation Boards. 'Scenery' belonged to a discourse in which natural beauty provided a desirable backdrop to Progress — the old notion of the pastoral in which nature existed primarily as a theatre of human events, not unlike one of those painted canvases in the style of naturalism that adorned the theatres of the day. The ALP Premier at the time, Eric Reece, proclaimed the new view on the site of Pedder to be just as 'scenic' as the old: 'Our engineers', he said, 'have changed contours with the aesthetic achievement of landscape gardeners.' Dams were not *merely* dams, they were monuments in a natural landscape; a new and technocratic model of pastoral beauty.

UNESCO, on the other hand, described the inundation as 'the greatest ecological tragedy since European settlement in Tasmania'.

In the campaign to make their case, the Save Lake Pedder activists advanced one of the earliest post-war critiques of large-scale industrialisation in general, and the claims of the Tasmanian Hydro-Electric Commission in particular. In the process they initiated a new and ecologically framed notion of the common wealth. Among their adherents were a significant number of professional experts including zoologists, physicists, geographers and engineers, all of whom knew enough science to use science against the technocrats. The environment movement never was anti-Rational and science was, and remains, one of the most potent weapons in the green movement's arsenal of polemics. At the time, however, the Pedder activist group was widely perceived as a single issue movement, and a rarefied one at that, and the success of their opposition lay in portraying them as middle-class purists who rated their own recreational privileges above the employment needs of the battler. In this way they were defeated both politically and morally: they could not carry the majority with them. In contrast, the Green Ban movement in urban Sydney was perceived as belonging to a program of broad social utility. It was much easier to enlist popular opinion in the support of saving Centennial Park than a remote alpine lake, although this it must be said was not the view of the leadership of the NSW Builders Labourers Federation that had imposed the Green Bans. In 1973 NSW BLF leader Jack Mundey visited Hobart to urge a 'blue ban' on the flooding of the lake but the Tasmanian Trades and Labour Council, at that time under right-wing control, blocked any such proposal.

In recognition that they had failed to get their message across, the Pedder activists resolved to contest elections and in March 1972 they formed the world's first Green party, the United Tasmania Group or UTG (the first national Green party, the New Zealand Values Party was formed in May of the same year). From the very beginning it was a broad-based coalition that defied orthodox categories. Its slogan – 'Politics of the Left and Right Versus Politics of the Future' was, as Richard Flanagan has observed, to anticipate the German *Die Grunen*'s 'Neither Left nor Right but in Front' by nearly a decade. The leader of the UTG at this time was Richard Jones, an academic biologist at the University of Tasmania. A forceful campaigner and a far-sighted thinker, Jones was to prove prescient in his belief that the Lake Pedder issue was an early warning sign of 'a fundamental breakdown in politics in Australia' and a growing disil lusionment with post-war consumerism and technocracy. It was Jones' belief that in many areas the old loyalties no longer applied and indeed the composition of the UTG showed it to be a mirror of emerging dissent that cut across traditional party lines, in this as in other ways a clear forerunner of the Greens Party today. Among its founders were former Liberal and Labor activists, while Jones himself had once been a member of the Queensland Country Party.

In May of 1972 the UTG fielded candidates in every Tasmanian state electorate. Despite hostile campaigns run by both parties and the local press, the new party polled a respectable 3.9 per cent of the overall vote (more than the Democrats in the WA state election in 2001) and narrowly missed winning a seat in one of the five multi member electorates. (At that time, under Tasmania's unique Hare Clark system of proportional representation with seven members per electorate, a candidate needed to poll 12.5 per cent of the vote to be elected in any given electorate.) In 1974 the UTG issued an economic, social and cultural manifesto set out in a document entitled 'The New Ethic'. This offered a critique of civil rights as well as current economic and social policy, rejecting not only 'the wholesale extraction of non-replenishable resources' but also the 'alienation of people in their social and work roles'. The northern Tasmanian newspaper *The Examiner*, in a rare show of dissent from the scare-mongering of the combined political establishment of that time, pronounced 'The New Ethic' the best set of policies on offer in the coming election.

For the next seven years, until 1979, the UTG contested several state and federal elections without winning a single seat, but in the course of these campaigns a new breed of political activist came into being. One of these was Bob Brown, the young doctor from Liffey who stood on the UTG Senate ticket in 1975 behind Dick Jones and gained only 112 votes – not that the experience was lost on Brown. 'Finding the United Tasmania Group', Brown would later write, 'was like being let out of the prison of conventional thinking.'

Certainly the rhetoric of Pedder was anything but conventional and there were times when the campaign was fought with a quasi-mystical fervour. That reverence translated into a new language that was to enter into Australian political discourse around this time, a discourse hitherto dominated by talk of 'workers' or 'battlers' and Menzies' own 'forgotten people'; of labour versus capital and the Cold War. But the environmental activists spoke a new language, one that in some respects harked back to the Romantic movement of the nineteenth century but in other and crucial respects went beyond it. On the Romantic side there was a certain religiosity of rhetoric. 'Material satisfactions are no substitute for spiritual content,' wrote Jones, 'and spiritual content will be denied any society which has lost the ability to live in harmony with its environment.' Another spokesman for the Lake Pedder Action Committee described its members as 'pilgrims' and wrote: 'I saw my temple ransacked by my own community.' Even a local politician was to describe his first sight of the lake as an experience akin to coming 'face-to-face with Jesus Christ for the first time'.

Unlike in the US, such publicly avowed sentiments were rare in Australian politics. It is true that for the previous fifty years much of

Australia's party political warfare had been conducted along Catholic-versus-Protestant lines but this was at least as much about cultural and tribal difference as it was about matters theological. When religious sentiment erupted within political discourse it was mostly in reference to the Cold War and the occasional invitation to the Divine to smite the Red Menace. But the green invocation of the spiritual was something new, one of the signs of an emerging political sensibility that had affinities with the cultural revolutions of the sixties and the phenomenon that a hitherto obscure Yale professor, Charles Reich, had chronicled in his surprise bestseller, *The Greening of America*. 'There is a revolution underway,' Reich had written, and 'it is not like the revolutions of the past. It has originated with the individual and with culture, and if it succeeds, it will change the political structure only as its final act. It will not require violence to succeed, and it cannot be successfully resisted by violence ...'

Thirteen years on, Reich's prophecy was to receive an emphatic demonstration on the banks of a wild river in Tasmania's south-west.

The Franklin River blockade of 1982–3 is one of the defining moments in Australian political history. It was the moment when ecological politics entered the mainstream. For the first time a well-organised group of environmental activists mounted a successful propaganda war in the media and carried public opinion on a national scale. For the first time a green organisation (The Wilderness Society) became a player in deciding the fate of the major parties – at both state and federal levels – while the theatre of the blockade itself became the crucible of a new generation of activists who came from all over Australia to take part.

In July 1976, in the aftermath of the dismal Pedder outcome, sixteen people came together at Bob Brown's old weatherboard cottage in the bush at Liffey just fifty kilometres outside the northern Tasmanian town of Launceston. This was a gathering of the South-West Tasmania Action Committee (SWTAC), set up in 1974 by Kevin Kiernan, one of the original Pedder activists. Out of this meeting came a new name, the Tasmanian Wilderness Society, later abbreviated to The Wilderness Society (TWS), and it was to be an inspired choice. Though Bob Brown had some misgivings about it at the time, to the extent that it might be interpreted as betokening a kind of remote elitism, the word 'wilderness' would quickly come to embody a passionate poetry that would prove galvanising. The 'Australian Conservation Foundation' sounded bureaucratic and responsible but 'wilderness' had a wild, inspirational ring to it that was a poetic clarion call, especially to the young.

By this time Brown had emerged as the leader of the Tasmanian environment movement and would soon, through the Franklin campaign, become its national face as well. In early 1976 his primary focus was the anti-nuclear campaign, but then he accepted an invitation from a fellow conservationist to go rafting down the Franklin River in the heart of Tasmania's south-western wilderness. Paddling around a bend in one of the river's steep, rocky gorges, he and his companion were shocked to discover

Hydro-Electric Commission (HEC) employees carrying out preliminary work on a proposed new dam. One minute, as he recalls it, 'magnificent canyons and waterfalls', the next 'jackhammers and explosions, helicopters – the works'. A series of three dams were to be built and two of the last remaining wild rivers in the world, the Gordon and the Franklin, would be flooded. Full-scale construction would commence in 1980.

After the Liffey meeting, The Wilderness Society went into action. Backed by funding from the Australian Conservation Foundation it embarked on a campaign to save the rivers. It was a campaign that would take far longer and prove more complex and exhausting than anyone could have foreseen at the time. It would encompass two stormy state elections, the destruction of a state government and a controversial state referendum on where to build the dam. It would also involve an unprecedented national effort that would culminate in the first federal election to be fought in part over an environmental issue.

Both the Tasmanian state Labor and Liberal parties were bent on building some kind of dam, though they were divided on where to build it. Early in 1982 the Fraser government agreed to nominate Tasmania's South-West for World Heritage Listing, but the Liberal Party was strongly federalist and Fraser was unwilling to contemplate intervention to overide state powers in the event of the proposed dam area being added to the list. The Wilderness Society lobbied politicians at all levels but made little headway. At the grass-roots level, however, it was a different story and seventy branches of The Wilderness Society sprang up around the country within two years of the formation of the original group. The newly formed Australian Democrats Party (1977) pledged to oppose the damming of the rivers and the first green activist to be elected to an Australian parliament, Dr Norm Sanders, entered the Tasmanian House of Assembly as a Democrat in 1980 on a wave of anti-dam feeling. At the same time the pro-environment lobby within the national ALP was growing in strength and at the party's biennial National Conference in Canberra in June 1982 a motion was passed by a clear majority pledging a federal Labor government to oppose the construction of a dam on the Gordon and Franklin Rivers. This meant that both the ALP and the Democrats would fight the next election on a platform that included a No Dams policy, a measure of the degree to which, in five years, The Wilderness Society had succeeded in making its case. Perhaps the most vivid emblem of its growing momentum was the famously romantic photograph of the Franklin River, *Rock Island Bend* (taken by wilderness photographer Peter Dombrovkis), which by then was appearing on posters and campaign material throughout the country.

By 1982 the Tasmanian Labor government had torn itself apart over the issue in what was perhaps the stormiest passage of government in the state's history. As the Tasmanian community continued to divide over the dam question, Labor was replaced by an even more aggressively pro-dam Liberal government which ordered an immediate commencement of work on the Franklin River. It was at this point that The Wilderness Society resorted to its contingency plan, an option it had been holding in reserve for some time. It would resort to a last-ditch blockade of the river itself, a full-scale campaign of civil disobedience that historian Tim Doyle has described as perhaps the most famous of all wilderness struggles.

The blockade began on 14 December 1982. Fortuitously it was the same day as news of the river's World Heritage listing came through. Over the next two months the blockade would become the first mass application in Australia of the techniques of civil disobedience invented by Gandhi many years before and refined by the US civil rights movement in the sixties. In preparation for it The Wilderness Society had flown in experts on Non-Violent Direct Action (NVDA) from New Zealand and training was held at the Baptist youth camp upstream from the Liffey River. Planning meetings had been set up in different houses around Hobart because the members of The Wilderness Society had evidence to suggest that their phones were being tapped. The NVDA method ran on a consensus model of decision-making based on a unit called the affinity group. Arrivals at the blockade reported to base camp on private land outside the small fishing town of Strahan on the state's north-west coast. There they were immediately assigned to a group of around twelve members and embarked on a two-day course of on-site training before being sent 'up river'. For many of them the blockade was to prove one of the formative experiences of their lives. Here is an account from a Sydney man now in his late forties:

> I drove up with my girlfriend to the main camp at Strahan and we registered in the dark outside this tent and then went off and pitched ours. Next morning we met up with about a dozen other people and this was the group we had been assigned to. These groups were called affinity groups and they were part of your basic training in the base camp. No one was allowed to go up-river without training. For three days we met for about four hours a day. It was a process of sitting around and talking and getting to know each other. The main idea was to develop trust so that when you were up river and confronted by trouble you wouldn't suddenly freak out and lose your nerve, or go the other way and do something stupid. We were mostly young, under thirty, and I remember base camp being full of young people who had come from all over Australia, taken time off work, leave without pay. Our trainer was a middle-aged guy called Jack Lomax who had this air of good-humoured calm. Jack explained the principles of non-violent action and I remember that we sat around in a circle and for some of that time we held hands. There was a lot of role-play and we took it in turns to play police and protesters, and simulated getting arrested.
>
> I remember that on the second day after I arrived there was some excitement because a group who were part of the way through their training transferred their 'class' to the wharf where police and HEC workers were attempting to load a bulldozer onto a barge. The group's training in consensus decision-making continued as its members sat down behind the moving vehicle while it began to

back up. 'Okay, how far do we want to go? Is there agreement that we sit here?' The heavily loaded truck rolled on. Blockaders were under the tray very near the wheels. 'Okay, are we still in agreement? Move back a bit, or stop? Stop? Okay. Any disagreement?' At this stage the police finally intervened and stopped the driver from squashing several members of the group who were still discussing whether or not they should stay.

After three days our own group went up river on a ferry and we were dropped off at an empty camp site where we pitched our tents on damp ground. Conditions were appalling, almost constant rain. In one of the storms lightning struck a tent and one guy had his jeans burnt from one leg. By this point we were already 'trespassing' in the wilderness. The next day we lined up on the river in rubber duckies to block a barge that was bringing a huge tractor up river to rip out trees. That pretty well ploughed through us and I remember having to push off from its side at the last minute. Then we paddled up river to Warner's Landing, went ashore and got arrested. After this we were taken back to the temporary magistrate's court set up in Strahan and given bail. Some wouldn't take bail and went to jail as a protest. The whole thing was pretty intimidating, and there were times when I was scared. Really I went in the first place for a bit of an adventure but in the end I got more than I bargained for. Once I actually saw the river, even in the middle of all that miserable weather … well, it's an experience very difficult to put into words.

Some of the eager arrivals at base camp had a different experience, however. They found the affinity group model confronting and left. An acquaintance of mine in the Labor Party lasted half a day in his group before packing up and driving back to Hobart. 'It was all that touchy-feely stuff,' he told me, grimacing with distaste. Touchy-feely was a long way from what young apparatchiks in the ALP were accustomed to.

By the end of February 1983 more than 1,400 blockaders had been arrested, 600 of them choosing to go to gaol. These were described by the Tasmanian Liberal Premier of the day, Robin Gray, as 'professional trouble-makers' imported by The Wilderness Society. 'Conservationists training for the south-west blockade are like guerrilla forces in third world countries,' he said. Several international figures came to be arrested (the naturalist David Bellamy prominent among them) and this ensured ongoing media coverage of a favourable character. Only two politicians were arrested, a Victorian Labor backbencher, David Gray, and the former Tasmanian Labor Minister for National Parks, Andrew Lohrey. Lohrey had played a vital role in behind-the-scenes cabinet manoeuvres to save the Franklin, assisted by a cohort of anonymous bureaucrats who had green sympathies. It was Lohrey's initiative that led to the creation of a submission for World Heritage Listing (subsequently endorsed by the federal government) and his insistence that saw the submission through many fraught cabinet meetings at state level until final agreement was reached.

Thanks partly to vivid footage of Bellamy and others on the evening news, polls showed public opinion to be solidly behind the blockade and in a significant reflection of what had been a rapid groundswell, Bob Brown was named the *Australian* newspaper's Australian of the Year for 1982. It was 1 January 1983 and Brown was in Risdon Prison. Shortly after his release he was seriously assaulted by four men outside Queenstown, one of them wielding a wheel brace. Twelve hours later he addressed a rally near the blockade site. By now he was a major media figure.

Jack Mundey has called the Franklin one of the most effective cases of mass action in Australian history and for a while the media couldn't get enough of it. The blockade was high drama and it ran well on the evening news. Unlike the blockade of a uranium mine — a dusty road, a wire fence and a drab gate a long way from the mine itself — the Franklin blockade was conducted on a 'scenic' site of wild and dramatic beauty. More importantly the river gave the media access right up to the point of the conflict. The weather was dismal, the terrain intractable and the police were not in control. Even getting arrested was dicey as the river would sweep blockaders in their rubber duckies past the point where they needed to land in order to 'trespass' and they would have to back-pedal furiously to counter the swirling current. Every night on the evening news it was the river itself that dominated the blockade, dwarfing both sides of the dispute. It was the river that caught the public imagination. The scenes on the river, wrote Tim Doyle, became 'a form of theatre – a passion play – its images entrenched in the psyches of all Australians'.

But despite winning the war on the symbolic level, The Wilderness Society was losing it on the ground. The blockade was a propaganda triumph but had proved ineffective in slowing the work of the HEC. Seeing this, the media sniffed defeat. The 'passion play' had all but run its course and still Fraser refused to intervene. The Wilderness Society was confronted by the dilemma of what to do next. And then, in the words of journalist and Franklin activist Peter Thompson, 'in one of those moves of fate that seems to have favoured the Greens throughout their history', Malcolm Fraser called an election.

Fraser, famously, had been motivated by a fear that the ALP would replace Hayden as leader with Hawke but for The Wilderness Society the timing was a godsend. It now decided to run a marginal seats campaign, to target seventeen key seats in all parts of Australia but particularly in metropolitan Sydney and Melbourne where voters were urged to 'put the Liberals last'. By this time the leaders of The Wilderness Society were battle-hardened and Brown warned the Liberals not to underestimate them: 'We are now old-timers in the political business,' he said. 'We have been through one federal, two state elections and a referendum in Tasmania. We are very hard-nosed in this political business.' Thirteen seats were selected for intensive campaigning and another seven for secondary attention. All twenty seats needed a swing of 3.8 per cent or less to be won by the ALP.

In all the drama of Hawke's sudden ascent to the leadership and his personal charisma that was made so much of at the time, the Franklin issue remains an under-acknowledged factor in Labor's 1983 win. Doug Anthony went so far as to say 'there is no doubt that the dam was the issue that lost the government the election' and he wasn't the only party numbers man to take the point. 'Failure to use federal power to stop the Franklin dam had cost Malcolm Fraser dearly,' Graham Richardson would later write in his memoirs, 'even though he had made some good decisions about Kakadu, the Great Barrier Reef and Fraser Island.' On the night of his victory Hawke announced that the new Labor government would use the federal powers granted by the World Heritage listing to intervene to prevent the flooding of the Franklin River. Beside him on the podium Hazel Hawke wore a pair of earrings embellished with The Wilderness Society campaign insignia of a green triangle enclosing the words 'NO DAMS'.

In one of those classic state-versus-federal stoushes that are so characteristic of the history of the federation, the Tasmanian state Liberal government immediately moved to challenge this intervention in the High Court. After all the drama of the blockade and months of intense lobbying behind the scenes it seemed that The Wilderness Society might yet lose. For several weeks a large number of Australian voters held their collective breath until, on 1 July 1983, the High Court ruled to uphold the Commonwealth's power. But it was a split decision – four to three – and two of the four majority judges, Brennan and Deane, had been appointed to the court in the eighteen months leading up to the case.

The question afterwards was this: why had a wild river in a remote and hostile landscape that most people had never seen nor ever would see captured the public imagination to such a degree? What did this notion of 'wilderness' stand for in the social imaginary? There are two possible answers to this. The first was that the river had become a symbol, potent at both the conscious and unconscious levels of the psyche, of the renewal of life itself. Some of this feeling was summed up at the time by the novelist James McQueen when he wrote: 'It seems to me now that the river – the River – is not just the Franklin, not just a river. For those of us who have been drawn, often despite ourselves, to its defence, it has become far more than the sum of its parts. For me it is the epitome of all the lost forests, all the submerged lakes, all the tamed rivers, all the extinguished species. It is threatened by the same mindless beast that has eaten our past, is eating our present, and threatens to eat our future: that civil beast of mean ambitions and broken promises and hedged bets and tawdry profits.'

There is a fascinating passage in the memoirs of Graham Richardson, later to become the most effective Minister for the Environment in any Australian government, that indirectly bears on this. A consistent hardliner in support of uranium mining, Richardson reflects on why, in his support of the Roxby Downs project, there was never any real concern in his mind about electoral disadvantage. 'Anyone who visits branches and electorate councils regularly, as I have done,' he wrote, 'can always tell when an issue really worries the rank and file. They never stop telling you about it: not just the hard-line activist cadres in the branches, but everybody.' Yet despite active public campaigns and the organised Left opposition within the Party that succeeded in containing the extent of the mining, 'the anti-uranium feeling from the rank and file simply wasn't there.' This was the feeling of many anti-uranium activists in the ALP at the time, though they wished it were not so. No anti-uranium campaign ever succeeded as the Franklin issue did in mobilising mass action that cut across class and party lines. And perhaps this had something to do with the colour green. Uranium was found in the desert, a landscape already characterised as a site of absence or lack. The river ran through one of the most fertile regions in the world. It became a symbol of love of country, something that on a large scale was consonant with the more modest but significant trend towards the re-imagining of the good old Australian backyard and the growing preference for planting bush flora in suburban gardens. 'A people were beginning to love their land,' wrote Peter Thompson.

Certainly it was true that for a whole new generation of young activists, two of their most formative political experiences were the Franklin blockade and their first visit overseas. One of these was the future Leader of the Greens parliamentary party in Tasmania, Christine Milne, who would later write of the impact on her of travelling in Europe as a young woman. 'I was stunned by the land degradation, acid rain and pollution, the loss of forests and polluted waterways of Europe. I was threatened by the nuclear power stations. After coming home through Asia, I was appalled by the affluence and wastefulness of the Australian lifestyle, as well as the inherent racism in our society. The imbalance in the consumption of the earth's resources became personal reality as well as known fact.' A similar response was put to me by a friend in that era. 'I'd been brought up on an idealised European landscape from children's storybooks. When I got there I was shocked. First, there were too many people in too little space. Secondly, there was this overwhelming sense of property, that every square inch of land was owned and had been owned many times over. Thirdly, there was all this pollution. When I saw Tasmania's south-west I realised that here in Australia we had something special on our own doorstep.'

With hindsight the Franklin River campaign can be interpreted as one of the emblematic outcomes of that new mood of Australian nationalism that emerged in the seventies with the Whitlam government and which included a radical re-visioning of country away from the idea of a 'dead heart' towards that of a 'red centre' and the notion of places like Uluru and the Franklin as sacred sites. It can be seen in the new wave of Australian cinema and films like Peter Weir's *Picnic at Hanging Rock* in which the plot seems merely an excuse to create a stunning meditation on the profound and enigmatic beauty of the bush. In *Picnic* it's as if a new generation of Australians is seeing its landscape for the first time, and this is reflected further in the work of artists like Fred Williams and Arthur Boyd as well as writers like Rodney Hall, Judith Wright and David Malouf. On the more domestic plane, traditional English gardens began to lose favour as, all over Australian suburbs, backyards were replanted with native trees. In the understated but moving conclusion to the David Williamson/Bruce Beresford film *Don's Party*, the young suburban hero, his hopes dashed at the end of a dismal election night, repairs to the front garden and consoles himself by tending his newly planted native saplings. It is a small moment but a telling one.

All political wars are waged in part on the symbolic plane (which perhaps accounts for why the Australian Republican Movement has so far failed to capture the public imagination) but there was more to the success of the Franklin campaign than a resurgent Australian nationalism accompanied by a new and potent emblem of the sublime. It is also true that here for the first time The Wilderness Society succeeded in broadening its political pitch. In this way it at last managed the shift, in the public mind, from a single issue party to something more. Traditional labourism and economism had been able to defeat the challenge to the Pedder flooding by combining to portray the Pedder activists as a fanatical minority who cared little for the common wealth. By the end of the Franklin blockade, that situation had changed, partly because the culture had changed but also because in order to save the Franklin The Wilderness Society had to do more than extol the wilderness, it had also to launch a comprehensive critique of modern industrial policy and the funding of the Hydro-Electric Commission. The United Tasmania Group had attempted this ten years earlier but by the time of the Franklin River dispute the HEC had over-reached itself and was in a more vulnerable position, especially given that there were now elements within the state Labor government who were pro-environment.

Although the HEC tried to keep the cost of its developments secret, in 1978 the then Minister for Energy, who was also Minister for the Environment, Andrew Lohrey, had revealed the massive blow-out of costs on the Pieman River scheme and the HEC's gross over-estimates of future power needs. This alone began to raise doubts in the public mind about the cost-effectiveness of hydro power as well as cast doubt on the legitimacy of the HEC's authority. A once respected institution could not be trusted. The Wilderness Society and its advisers were quick to capitalise on this and worked hard to develop an alternative economic policy. The money squandered on the Pieman, they argued, could be better spent on schools, hospitals and roads, or on developing small business: the government was spending huge amounts of public money to prop up foreign-owned big businesses that were in fact reducing their workforces. It was at this point that green activists made their first effective claim on the traditional utopian narrative that had hitherto been the domain of Labor: what is the most equitable distribution of resources?

By 1983 many voters who hadn't the least interest in wilderness opposed the damming of the Franklin River in the belief that massive public expenditure on dams was economically unjustifiable. The electoral support of environmental candidates broadened. Greenies were no longer bearded backpackers and vegetarian cranks: they were becoming assimilated into the old utilitarian notion of the greatest good for the greatest number. Within just a few years The Wilderness Society had succeeded, in Mungo MacCallum's words, in turning 'a minority cause into a mass movement'. A new political sensibility had broken through into the mainstream.

In 1984 The Wilderness Society emerged from the Franklin campaign with a national organisation and enhanced credibility – which was just as well because the forests campaigns lay ahead of it. What it couldn't foresee was that the environment movement was about to acquire a new and unexpected champion within the Hawke Labor government with the appointment in 1987 of Senator Graham Richardson as Minister for the Environment. It was partly because of Richardson's strategic electoral instincts, not to mention the personal rapport he appears to have struck with some of the movement's leaders, that substantial green gains were to be made over the next few years.

According to Richardson in his memoirs, *Whatever It Takes*, it was clear to him from the time of the Franklin blockade that one of the things it would 'take' was for Labor to show more leadership on the environment if it was to carry the marginal seats. The warning signs to the ALP of what might happen in the long term if a well-organised Green party got going are there throughout Richardson's autobiography, accompanied by an absorbing account of his own ongoing strategy to pre-empt any such eventuality. In a chapter entitled 'Out of the Smoke-filled Rooms and into the Forests' Richardson describes the seminal occasion in 1986 (before he was appointed Minister) when The Wilderness Society invited him to fly over Tasmanian forests. 'By the time we arrived back in Hobart,' wrote Richardson, 'I was a convert. Having been shown the awesome forests and streams he [Brown] wanted protected, I wanted to become a warrior for his cause … It didn't take too long to work out that we had a perfect convergence: what was right was also popular.'

Richardson also speaks of Philip Toyne, then executive director of the Australian Conservation Foundation, as 'another character I found difficult to refuse. Like Brown, Toyne had the sort of commitment to the cause that would never let him give up and admit defeat.' Richardson arranged for Brown and Toyne to see Hawke about the issue of Tasmanian forests as well as World Heritage Listing for the wet tropical rainforests of north Queensland and some clear demarcation of the expanded boundaries of Kakadu National Park. At this time Kakadu, not forests, was becoming the symbolic issue in the public mind and the ongoing effect of the Franklin blockade can be gauged by Richardson's remark, 'I knew that if we didn't look after Kakadu it could become our Franklin River.' Hawke needed no persuading. He knew instinctively, wrote Richardson, 'that the ACF and the Tasmanian Wilderness Society were [Labor's] natural allies and that they had to be brought into the tent. He had embarked on a green course that would do him no end of electoral good in the next few years.'

Labor's green strategy paid off. In the close election of 1987 the ALP won on preferences with the green vote worth a crucial half to one per cent. The broad-based and informal alliance of community groups that was coming increasingly to be known as 'the greens' was confirmed as one of the most influential lobbies in the community though they were not yet a party in their own right. The record of the Democrats on environmental issues was strong and many green activists were happy to support them as their de facto political arm. At the same time, other greens saw themselves as more radical than the Democrats and were keen to pursue the European model of direct political representation. In 1983 the Sydney Greens had registered as the first local Green party and in 1984 on her second visit to Australia the German Green parliamentarian Petra Kelly urged the leaders of the various environmental groupings to forge a national identity. A year later, another state-based party, the Queensland Greens, was established but for the moment the national focus, in regard to small parties at least, was the formation in 1984 of the Nuclear Disarmament Party (NDP). At the head of an NDP ticket, Jo Vallentine was elected to the Senate from Western Australia with Peter Garrett narrowly missing out in New South Wales.

Partly in response to Kelly's visit, a national Getting Together Conference was held at Sydney University in 1986, organised by Bob Brown, Brisbane activist Drew Hutton and other green leaders. Although over

500 activists attended, no agreement could be arrived at to set up a national party. On the one hand the small groups involved seemed too diverse (and on some points too ideologically disparate) while on the other many green activists felt that they had the ear of the federal government under Hawke. Still, the issue of a national party continued to simmer away and in 1987 Bob Brown approached Don Chipp in Canberra to initiate talks on how the Democrats, the Nuclear Disarmament Party and the Tasmanian groups might combine to form a national Green party. Not surprisingly – the Democrats having the upper hand at this stage – the proposal was rejected outright. As a result the greens continued to develop along state lines and in 1989 the South Australian Greens party was formed. Then, in 1990 four groups in Western Australia combined to form the Greens WA. When Jo Vallentine was re-elected to the Senate in March of 1990 she became the first member of the Australian Parliament to represent a party with the official label 'Green'.

Meanwhile, in Tasmania, a new battle was looming. In 1987 the logging company North Broken Hill in partnership with the Canadian mining and logging giant Noranda announced that they would build a 'world-class' pulp mill in the farming district of Wesley Vale on Tasmania's north coast. At this point it became clear just how much the Franklin River campaign had succeeded in making inroads into mainstream thinking because the announcement provoked an immediate outcry. The mill would be built on some of the richest farming soil in Australia and would pump thirteen tonnes of toxic organochlorine wastes into Bass Strait each day while blowing carcinogenic dioxins across surrounding farmland. The campaign of opposition that ensued is a dramatic illustration of Tim Doyle's palimpsest model in which up to thirty diverse organisations may come together to play a part. The Wesley Vale campaign produced an unprecedentedly broad alignment of interest groups with the Farmers and Graziers and the Fisherman's Association – groups not hitherto known for their radicalism – lined up behind CROPS

(Concerned Residents Opposed to Pulp Mill Siting), an action group organised and led by local schoolteacher and farmer's daughter Christine Milne. Large numbers turned out to demonstrate and for the first time environment marches were to feature a procession of farmers on tractors. Once again the state Liberal government threw its weight firmly behind the multinationals and announced that it would recall parliament in order to pass special laws enabling the mill to be built. The media release announcing this appeared on North Broken Hill letterhead.

At first Richardson supported the pulp mill proposal because it offered an advance on woodchipping. 'I took the view – and still do – that woodchipping is a scandalous waste of forest resources,' he later wrote. 'We export both the resources and jobs to Japan and are paid precious little.' But when it became clear that Noranda was not prepared to offer adequate environmental safeguards Richardson changed his mind. When the federal government imposed some extra, though modest, environmental standards with which the mill would have to comply, Richardson received an Environmental Impact Statement document from Noranda which he later memorably described as the equivalent of 'up yours'. The main game for Noranda was its investments in Canada, and knowing that the opponents of its polluting mills back home might call for similar safeguards, Noranda withdrew. The project collapsed and the greens under Milne's leadership had won a significant victory.

At the time the Wesley Vale case seemed to mark yet another expansion of the parameters of the local environment debate, out from beyond wilderness to a broader concern with everyday life. On a roll from the gains made in public opinion the greens mounted a strong team in the Tasmanian state election of 1989. Demonstrating that they had learned as well as anyone to appeal to self-interest and the hip pocket nerve (not least the expanding market in Europe for 'clean green' produce), this time they campaigned on a strongly instrumentalist platform and one heard little of the deep ecologists or the 'wilderness as cathedral of the soul' school. Their success can be gauged by the fact that, under Tasmania's system of proportional representation, the Green independents won five out of thirty-five seats in the lower house with neither of the major parties winning a clear majority. At first the defeated Liberal government refused to go and, in an effort to protect business and forestry interests, the newspaper proprietor and leading businessman Edmund Rouse attempted to bribe an ALP backbencher to cross the floor (Rouse was subsequently gaoled). When that manoeuvre failed the Green independents negotiated support for the minority Labor government in an unprecedented arrangement that became known as the Green–Labor Accord. Just seventeen years after the foundation of the United Tasmania Group, the Greens had the balance of power. They had averaged 17.1 per cent across the state and Bob Brown, with a personal vote of 23.5 per cent, had polled second only to the Liberal premier Robin Gray in state-wide primary votes. Exit polls showed that defections to the Greens had cut across traditional class and economic lines. In the south of the state there was a significant blue-collar vote while in the north there was a shift away from the Liberals among small farmers worried about the despoliation of agricultural land. Milne was elected in what was largely a rural seat and had office-bearing members of the Liberal Party working on her campaign (in some ways a sign of what was to develop into the Liberals for Forests phenomenon in Western Australia in 2001). It was an entertaining sight to watch the evening television news where middle-aged women in pearls and coiffed grey hair spoke vehemently on a platform they had once regarded as the domain of bearded cranks. Just before the election the Hobart *Mercury* reported a small farmer in the north as saying: 'A few years ago I would happily have run over Bob Brown with my bulldozer. Now if I saw him up a tree I'd take his lunch up to him.'

In light of the many European instances of Greens entering into coalition governments, it's worth noting here that in Australia's first Green-influenced government the Tasmanian Green independents were not prepared to emulate that model. The 1989–91 period was a Labor minority government. The Accord was a set of policy objectives that the Greens and the Labor Party signed off on. It was the mechanism for Labor to form government and be assured of a majority on the floor of the House of Assembly. The Greens in turn agreed to guarantee supply and not to support no confidence motions put forward by other parties but they reserved the right to move their own (in relation to matters such as freedom of information legislation). At no stage did any of the Green members enter Cabinet or hold ministries.

Just months after the signing of the Green–Labor Accord, Hawke called a federal election for 1990. By this time opinion polls were showing that the environment had come from twelfth to second place as an issue of public concern and Richardson briefed the ALP's campaign directors accordingly. The outcome spoke for itself and 1990 became known as the green election. Despite polling only 39.4 per cent of the primary vote Labor was returned by what Richardson described as 'an unbelievable preference drift'. In no less than ten seats, Labor won after being behind on primary votes because an overwhelming percentage of Greens prefer-ences went to the ALP along with a two-to-one majority of Democrats preferences. It was the highest proportion of preferences the ALP had ever achieved.

Significantly, between 1987 and 1990 the Democrats vote almost dou-bled. The green element of this was clearly a factor, as well as a growing disaffection among both Liberal and Labor voters with economic ratio-nalism. Once again the Greens approached the Democrats. In 1991 four of the Tasmanian Green parliamentarians met with Janet Powell and national Democrats officials to set in train moves towards what might have become the 'Green Democrats'. Not everyone in the state Green par-ties was in favour of this move but at that time there was no organised national structure to deliberate on a co-ordinated national strategy. Soon afterwards, however, Powell lost the leadership of the Democrats and her successor, Cheryl Kernot, indicated in no uncertain terms that she had no interest in a merger. Kernot had a different strategy in mind for the

Democrats, one that was to play out with some drama a few years down the track.

By this time the Tasmanian Accord had foundered. Opposition from the more conservative unions was intense as it was from business interests, especially from the logging and mining industries that traditionally were as close to Labor as to the Liberals. Newspaper editorials railed against the evils of minority government, conveniently overlooking the fact that minority governments were common throughout Western Europe. The conventional wisdom in Tasmania soon became that the Accord had been an unqualified disaster when in fact it represented one of the most progressive periods in the state's post-war history, however brief. The Tasmanian Greens achieved some major concessions, including a doubling in the size of the South Western Tasmanian World Heritage Wilderness, but irretrievable breakdown was reached after just fifteen months when the ALP abandoned the agreed-upon limit to the woodchipping of native forests and brought in Resource Security legislation for the major logging companies. Labor and the media declared the Accord a failure, implying that all minority government must lead inevitably to disaster, but at least one Labor ministerial staffer disagreed, citing an absence of goodwill towards the Greens from the start. Milne, who was later to succeed Bob Brown as parliamentary leader of the Green independents, claims that within weeks of the Accord being signed the state secretary of the ALP was freely announcing that it was Labor's objective to 'destroy the Greens'. She also holds the view that the Parliamentary Labor Party of the day was not strong enough to stand up to business interests and the right-wing unions led by Jim Bacon and Paul Lennon (now Premier and Deputy Premier in the current Tasmanian government under which woodchipping has reached record levels). 'A Graham Richardson type would not have fallen on his sword for the timber industry over Resource Security legislation,' says Milne, who worked closely with Graham Richardson over the Wesley Vale pulp mill proposal. 'It would have been dropped and not revisited.'

In the ensuing election, Labor lost office.

The Green leaders, meanwhile, continued to discuss the prospect of a national party. Bob Brown and Drew Hutton, especially, were keen to see if they could become something more than a network of like minds operating in decentralised groups, and eventually Hutton, Brown, Vallentine, Garrett and others in the conservation movement entered into a round of consultations and set up another conference in 1991. According to Hutton, this was a conference dominated by four main groups, each with its own tendencies: the nationalists who wanted an organisation with strong local branches but a national body as well; the decentralised, do-it-at-the-grass-roots movement that predominated in New South Wales; the Socialist Workers Party (SWP) that had made headway in the green coalition, especially in New South Wales; and the Western Australians who, even here, embodied the traditional Western Australian outlook that they belonged to another country. By this time The Wilderness Society, though still active, was not a key player. The environmental groups were still, in Hutton's words, the 'fire in the belly' of the green political movement but by now there were significant players from the peace movement, the NDP and social justice and civil liberties groups. Hutton himself, close to Brown on many issues, came out of the Queensland peace movement and civil liberties campaigns.

Eventually, after much discussion, a rule was introduced that members of Green parties could not be members of any other party, one of several measures that effectively disposed of the Socialist Workers Party influence. Purged of the SWP, New South Wales came on board and another conference was held in Sydney in late 1992, a formal conference with delegates elected from each state. Twenty years on from the foundation of the UTG, the Greens at last became a national organisation. On August 1992, in North Sydney, a press conference was held to announce the formation of the Australian Greens Party. Not one television news crew turned up to cover the event.

In hindsight this absence of media interest foreshadowed a change in the political mood of the coming decade. Looking back now, many of the

Greens see the nineties as a period of relative stagnation. Keating replaced Hawke and showed little sign of interest in the environment. Though in his memoirs Richardson portrays Keating as something of a closet green sympathiser, Keating had other agendas to run as prime minister. The 1993 election was dominated by bitter division over the GST as well as a perceived threat to the public health system. Still, in the two states where a green culture had been established for some time, Tasmania and Western Australia, the Greens performed respectably. Dee Margetts won a second Senate seat for the Greens WA while Judy Henderson was almost elected in Tasmania. Green preferences flowed to Labor in enough marginal seats to help ensure a narrow victory for Keating but the mood of the government had changed.

Under Hawke and Richardson, federal Labor had set up Environmentally Sustainable Development groups made up of representatives from industry, the conservation movement and government. These were working parties aimed at developing an industry policy that was integrated and long term, and not about one-off, hit-or-miss campaigns for individual bits of wilderness. Under Keating, however, the groups became ineffectual. There was a general cut-back in funding, many state governments were opposed to them and industry began to mobilise more effectively, either in terms of out-lobbying the conservation groups or in 'greenwashing' their development strategies. The big companies, especially in mining and woodchipping, showed that they had learned from the environment movement, even to the extent of setting up what the Greens claimed were front groups such as the Forest Protection Society and the Mothers Opposing Pollution. Keating set up the Regional Forestry Agreements, an arrangement within which the conservation movement, in the perception of many frontliners, got 'done over' by successive state and federal governments. Bob Brown has since described the nineties as a drought. 'The big parties and the corporate sector discovered greenwash, and a billion dollars of Telstra funds was going to save Australia's environment. It reassured the public for a while.' After the failure of the Green–Labor Accord, Brown resigned from the Tasmanian Parliament in 1993 and took three years off from representative politics. Meanwhile Labor continued to assume it had the green vote in its pocket. Where else could it go?

Drew Hutton points to other factors in the environment movement losing momentum in the nineties. For one thing it had become more professionalised. 'When you go setting up a whole lot of green bureaucrats in different environment centres or heading different organisations, these are wonderful, committed people but they're wonderful, committed people sitting on committees with government and industry, lobbying governments on a daily basis and doing everything except going out and mobilising supporters. And when push comes to shove and the industry is in there and you're getting frozen out, you've got to go back to your grass roots, you've got to go back to your constituency and mobilise them. If you can't do that, you get done over.' Some green supporters became complacent. 'Our supporters thought, well this is good, they're in there talking to government, we don't have to do so much.'

In the nineties, too, economics had replaced science as the official state religion. Suddenly stockmarket indexes were routinely reviewed on the evening news, usually just before the weather as if both were natural phenomena. Investment advisors had regular talk-back radio segments and by the end of the eighties junk bond traders like Michael Milken had replaced nuclear physicists like Edward Teller as the new Masters of the Universe – from *Dr. Strangelove* to *The Bonfire of the Vanities*. Courses in business management had been introduced into schools in the early eighties by proponents who would have denounced comparable training in trade union issues as untenable. Groups of high-school students acquired their own share portfolios and were written up in the media as good citizens in training. The phenomenon of the so-called Mum-and-Dad sharehold ers, later to transmogrify into the more phantom 'aspirational voter', became prosaic emblems of the end of ideology, everyday indications that the intellectual Right had won the ideological war in both of the major parties. Mrs Thatcher's dictum that there was no such thing as 'society', only an aggregation of self-interested individuals, was to be promulgated in a less flagrant and more astute version by John Howard, i.e. coated in a reassuring patina of old-style Protestant values that had little to do with the realities of the information age and the ruthless transformations of globalisation.

The Green–Labor Accord in Tasmania and the 1990 federal election began to look like peaks in a golden era that was over. It seemed that the green movement might have reached a plateau and maybe even stalled.

Despite this, the hard-core green vote continued to hold and the Greens were not without electoral gains in the nineties. They achieved a balance of power in the ACT state elections in 1995 and long-time peace activist Ian Cohen was elected to the NSW upper house in the same year. Then, in the Queensland state election of July 1995, the Queensland Greens orchestrated a campaign to upset Labor's complacency. When Goss showed little interest in environmental issues, for the first time they directed preferences away from the ALP. Goss made concessions and the trend was reversed. In a big swing against the government, Goss survived on Greens preferences to cling to a one-seat majority. The Greens scored 8.9 per cent on average in all seats contested and 24 per cent in Drew Hutton's inner-city Brisbane electorate. It was a hard-headed demonstration that Greens preferences could no longer be taken for granted. Meanwhile in Tasmania, after a period of Liberal majority government, the Green independents won back the balance of power and agreed to give support to a Liberal minority government since the Tasmanian Labor Party, carrying a legacy of bitterness from the Accord, declared it would rather go into opposition than into a Green-supported government. It remains the only Australian instance of a party faced with a choice of government choosing opposition (one can only imagine what Graham 'whatever it takes' Richardson thought of that). Despite tensions between the Greens and the Liberals, the next two years saw a productive period of government which delivered gun control, gay law reform, an apology to the stolen generation and a lower house vote in favour of a republic. By 1998 there were young Tasmanian voters coming onto the rolls who had never known a political system in which the Greens were not represented.

But the second half of the nineties was to be dominated by the rise of another minor party, One Nation, and another kind of 'love of country', one that had its own grievances in relation to the onward march of globalisation. On the national scene this was a dry period for the Greens, relieved only by the election of Bob Brown as a Senator for Tasmania in 1996. In 1998 the Greens, who had hoped to win a Senate seat in one of their strongholds, Queensland, received only 2.1 per cent of the vote. At the same time, while traditional labourism and class-based politics had gone into decline and economic rationalism in the guise of globalisation had won over both of the major parties, this did not mean that there was not considerable unhappiness in the community at this putative 'end of ideology'. In its reactionary mode the protest vote gravitated to One Nation while in its progressive form it manifested itself in the success of the Democrats who continued to do well as a broad-based protest party, so much so that they came to hold the balance of power in the Senate.

At this point Cheryl Kernot altered the strategy of the Democrats. Since their inception the Democrats had skilfully set about building a constituency: small business that was conservative on economic grounds but progressive on social issues (the 'wets' of the Liberal constituency); Labor people opposed to economic rationalism and the deregulation brought on by Hawke and Keating; and a soft-core green vote that switched to the Democrats in the late eighties and stayed with them. Under Janine Haines and Janet Powell the party had been oppositionist in tendency but Kernot embarked on a strategy (persisted in by Meg Lees) of moving closer to government in order to have a greater influence on policy. In this way the Democrats expected to enhance their appeal to strategic voters by demonstrating that they were not mere nay sayers but could make a constructive contribution to the parliament. At first the strategy appeared to work, but inevitably it had the effect of making the Democrats look more conservative. Even at the height of her success, Bob Brown had his doubts about Lees' ongoing strategy. Drew Hutton recalls a conversation in which Brown said to him, in effect: 'The Democrats are going to destroy themselves because their tactic is wrong. If you're going to hang in close to the Liberal Party and mitigate their policies, rather than hang in close and mitigate Labor's, you're actually starting from such a low base on things like the GST or industrial relations that you're going to get yourself into trouble.'

'It took a while,' adds Hutton, 'but they did.'

Up until this point Hutton had been in favour of some kind of alliance with the Democrats but, from around 1996 on, 'I realised we had only one choice and that was to replace them. Don't get me wrong,' he adds, 'I think there's a legitimate role to be played by having the balance of power. Strategic voters are people who want you to get things done. They don't elect you to play a symbolic role in parliament and vote against everything that turns up, they want you to play a clever role. But you have to do this in such a way as to keep faith with your constituency.' Brown agreed. He had already taken the controversial and hard-headed step of running, in effect, against a sitting Democrats Senator, Robert Bell, in the 1996 federal election. Bell was widely respected and well liked but progressive Tasmanian voters showed just how strategic they could be in switching their vote away from Bell to Brown (Brown was elected). I recall many heated discussions in Tasmania at the time about whether to abandon Bell but already there was a mood among green supporters that the Democrats were not doing enough. There was also the perception that it was not often that voters got the chance to put a politician of Brown's calibre into the national parliament. If anyone could make an impact it would be him.

The so-called dry years between 1995 and 2001 were potentially a demoralising time but the membership of the Greens Party held. 'We always felt the support was there,' says Hutton. 'Any other party would have folded. We were getting nowhere. We had Bob, and after '96 I think

we would have survived without Bob but we would have struggled, we'd
have been smaller than otherwise – but the pendulum would have swung
back because the material base was there.' The salinity crisis, greenhouse
emissions and the destruction of old growth forests were not problems
that were about to go away. This was something that astute reader of the
political wind, Graham Richardson, had foreseen way back in the mid-
eighties. 'The environment will always return as an issue,' he wrote,
'because it has to.'

> On winter nights we would always burn mallee roots, which the
> woodman regularly heaped on the 'nature strip' outside our front
> gate … In late summer if there was a hot north wind we might get
> 'red rain', big terracotta teardrops on all the picture windows and
> the car windscreens, which took hours to remove the next day with
> a chamois. The rain, we were told, was from the Mallee, a horrible
> far-away place where the firewood came from. It never occurred to
> me, or to many others then, that the more these roots were ripped
> from the soil to make our home cosy, the more red rain would
> freckle our nice clean windows. It was a tasteless reminder of a
> place where Aborigines had once lived, where explorers had died,
> and where none of us need ever set foot.
>
> – Barry Humphries, *More Please* (1992)

In one of the much-loved classics of Australian fiction, *My Brother Jack*
(1964), George Johnston captured a moment in the suburban philistin-
ism of post-war Australia that resonated for a great many of his readers.
Johnston's hero and alter-ego, David Meredith, is a hard-drinking, hard-
living journalist who suffers from an alienating sense of the soullessness
of his world. Prosperous and successful, he is nevertheless unhappy.
Settled into a new and affluent Melbourne housing development he
climbs one evening onto the red-tiled roof of his house and stares out
over 'the whole of the sterile desolation'.

> There was not one tree on the whole estate.
>
> Yet there must have been trees once, I thought … and now there
> was nothing but a great red scab grown over the wounds the bull-
> dozers had made, and not a single tree remaining …

Here Johnston is remarking on one of the ugly peculiarities of Australian
suburbia in the early postwar period, an apparent hatred of trees. It was

not only that it was easier for developers to flatten the landscape with speed; even settled suburbanites would quarrel over the large trees in their midst. The roots were held to damage the foundations while the foliage shed leaves and nuts that littered the lawns and clogged up guttering and drains. In short, trees created untidiness. Always, it seemed, there were issues around whether to root them out, cut them back or let them spread, and decisions either way led not infrequently to violence or litigation. Right up until the seventies the suburban gardener's great obsession was with neatness and ordered respectability; with pruning, clipping, lopping, raking, sweeping and hosing down. This meant that often, in a sun-struck landscape, living areas were bereft of shade. In the seventies, however, as part of that general shift in sensibility that Charles Reich set out to map in North America, the suburbs began to change. Local councils instituted tree-planting campaigns and gave seedlings away free. UNESCO introduced the International Year of the Tree and schools celebrated with annual tree-planting days. Critiques of land clearing and salinity damage began to get a wider hearing.

For a long time, however, the term 'tree hugger' remained an epithet of scorn, or at best a wry acknowledgement of a form of sentimentalism that had been allowed to get out of control. It was one thing to bring a domesticated version of the bush ethos into the backyard, it was another to make excessive gestures like chaining yourself to a tree, and the prevailing image of anti-logging protestors remained resolutely feral. Real men – men like George Johnston – didn't engage in arboreal melodramas. So adverse was media coverage of the 'dole bludgers and dreadlocks' scene that in 1992 Geoff Law, then campaign co-ordinator for The Wilderness Society in Tasmania, decided that the Society should abandon the style of direct action for which it was best known and work by other means. (In the decade since, only one small instance of on-site direct action has been mounted in Tasmania.) But media images can be deceptive on more than one level. There is the surface level to which Geoff Law took exception, but there are also other levels of reality which are more difficult for the observer to gauge. Even the most conscientious journalists cannot be expected to produce documentary evidence of subtle shifts of feeling that are occurring incrementally and often unconsciously in the hearts and minds of individual citizens. This is a level of reality that the American critic Fredric Jameson refers to as 'the political unconscious', and Jameson – a Marxist – has in mind a model of social relations arising out of a strong, even visceral desire for human community. Eco-philosophers have something else in mind, however. Alan Drengson, for example, posits the existence of a 'latent ecological unconscious' and Edmund O. Wilson in his famous biophilia thesis argues for an innately emotional affiliation of human beings to other living organisms, an affiliation that is always there, even when not apparent. Certainly there are moments when these propositions seem to be instantiated before one's eyes, as on the recent occasion when thousands of foreshore residents and office-workers rushed excitedly to the inner edges of Sydney Harbour to watch the spectacle of a sperm whale playing with her cub. And who among us doesn't feel a sense of awe, and loss, when we read accounts by the early settlers of how difficult it could be to cross Hobart's Derwent River because of the numbers of whales frequenting its waters?

So perhaps the pundits should not have been as unprepared as they appeared to be by events in Western Australia, where the relentless devastation produced by clear-felling of old growth forests culminated in an outburst of revulsion, at once aesthetic and moral, that seemed somehow to reach its defining moment in the middle of 1999 when an elderly matron from Perth climbed onto a tree stump and declared that enough was enough. This was no less a figure than Dame Rachael Clelland, one of the founders of the WA Liberal Party. Dame Rachael was but one of several disgruntled Liberals outraged by what they saw as the scandal of the Court government's Regional Forest Agreement. So strong were their feelings on the issue that they were prepared to sever their longstanding ties with the state Liberal Party and launch a unique breakaway group.

Looking back, it's no surprise that the Liberals for Forests phenomenon should have arisen in Western Australia. Firstly, it was the state where for a hundred years development had been at its most rapacious. Western Australia boasted the greatest degree of land clearing in Australia and had the highest rate of emission of greenhouse gases in the world. Secondly, and in response to this, the Western Australian environment movement was one of the oldest and most active in the country, with the best record of electing Green politicians to Canberra. It wasn't as if the outcry over the Court government's Regional Forest Agreement had come from nowhere; it had been preceded by a two-year anti-logging campaign involving a number of high-profile identities such as fashion designer Liz Davenport and basketballer Luc Longley, and in July 1999 divisions within the Liberal Party finally erupted into the launch of a breakaway group led by the former federal president of the Australian Medical Association, cardiologist Keith Woollard. Woollard began by announcing that independent polling showed 80 per cent of Coalition voters were opposed to any further logging in old growth forests. Here was a dying industry, he declared, that was heavily subsidised and destructive and which exploited a natural resource over which governments had absolute control. There was no way that it made sense 'according to conservative economic principles'.

At the launch of Liberals for Forests there were two former Vice-Presidents of the State Liberal Party, two state independent MPs and Dame Rachael Clelland, who was quoted in the press as saying: 'This issue is quite beyond party politics. It's a mass movement in Western Australia to keep these forests. The issue is almost unanimous amongst the people … people stop me in the street. The check-out girl in Woolworths, they're all the same. "Thank you for what you're doing to save our forests."' Even the leader of the WA National Party, Hendy Cowan, broke ranks with the government. When the ABC's Terry Lane interviewed Cowan on 'The National Interest' in the week leading up to the 2001 state election, he paused in the middle of the interview to say, 'And if you've just tuned in to the program and can't believe your ears, yes, this is the Leader of the West Australian National Party sticking up for old growth forests.' Meanwhile the Labor Party responded by running harder on an environmental platform than any party in any state election ever, pledging to preserve 99 per cent of remaining old growth forest.

As in the Wesley Vale campaign in Tasmania more than ten years before, the green political landscape began to exhibit new accoutrements. A group known as Suits for Forests held a rally outside Premier Richard Court's city office and, using their mobile phones, number-jammed his call lines. When I heard a report of this rally on the news I knew that Court had lost. It was a sure sign that a shift in sensibility had taken place. I'd lived in Tasmania and I knew the signs. 'You slog away for years,' as one green activist reported, 'and you feel we're getting nowhere on this but what you've been doing for twenty years is inculcating a deeply felt attitude amongst people, and at the right historic moment this will come out – this is too much, this is enough. And we saw that the penny had finally dropped.'

But the penny hadn't dropped for everyone. In the *West Australian* newspaper the day before the election a panel of experts predicted a Liberal victory. Instead, with the aid of Greens and One Nation preferences, Labor won by a comfortable margin of seven seats. (One Nation had refused to give preferences to any sitting members which, since Labor had fewer seats, worked against the government.) The three sitting Greens members in the upper house who were elected in 1996 were returned along with two more. These now hold the balance of power. In the eight seats contested by Liberals for Forests the group secured on average 10.6 per cent of the vote. One of its candidates, Dr Janet Woollard, was elected to the lower house while Liz Davenport came within an electoral inch of unseating Court in his own safe seat of Nedlands. In some seats the combined vote of anti-logging parties was as high as 30 per cent. One Nation managed 9.87 per cent, down slightly on its 1998 result, and in the eight seats where One Nation faced both Greens and Liberals for

Forests, its primary vote totalled 12,156 compared with the combined environment vote of 28,174. Preferences from the Greens, Liberals for Forests and independent environment candidates handed Labor three of these eight seats. All three had been held by Liberal ministers. As for the Greens WA, they had jumped two points higher than the Senate election in 1998 to poll 8 per cent. Despite some expensive newspaper advertising featuring Meg Lees, the primary vote for the Democrats dropped to 3.6 per cent and they failed to win a seat, both sitting members in the upper house losing their places.

The media was quick to identify signs of a green resurgence. A report in the *Australian Financial Review* proclaimed 'Green politics is back'. (16/2/01, Chelsey Martin) 'Ferals climb into bed with the suits' reported the *Sydney Morning Herald* (3/2/01, David Reardon), and 'greenies rub shoulders with the top end of town'. The *Herald* also proclaimed that green was 'fashionable again', but the most telling article was one tucked away in the bottom corner of the *West Australian* (12/2/01), 'Voters couldn't see the forest for the trees'. This concerned the former Democrats members who had lost their seats in Western Australia's Legislative Council, Helen Hodgson and Norm Kelly. Yes, the GST was partly to blame, said Hodgson, but more to the point, perhaps, was the fact that the Democrats Party had appeared to suffer from a 'protest vote' against it because instead of being seen as the biggest minority party it was now perceived as the smallest of the major parties. In other words, once the protest vote had been for the Democrats, now it was against them.

Hodgson's ominous perception appeared to be borne out soon after in the June 2001 by-election for Richard Court's seat of Nedlands. While the Liberals managed in the end to retain the seat, they had to fight off a challenge from the Greens to do it. Despite the newly elected Labor government taking rapid steps to implement its anti-logging policy, the Greens increased their primary vote to 14 per cent. Liberals for Forests polled 12 per cent with the majority of their preferences flowing to the Greens. Labor polled 18 per cent and the Democrats 6 per cent. The fact that the Greens had done even better than before prompted the new leader of the Liberals, Colin Barnett, to warn the Howard government that it must make the environment a higher priority. For those with their ears to the ground it was clear that 'Liberals for Forests' had not been only about trees. It had become code for all those people who were fed up with the increasing soullessness of their party.

In the two years leading up to the election several celebrities had spoken out against the clear-felling of old growth forests but perhaps none had been more influential, or more noteworthy, than the coach of the West Coast Eagles, Mick Malthouse. Long before the 2001 election Malthouse had made strong public statements in opposition to clear-felling, attracting hate mail and death threats for his trouble. Against the wishes of his club management he had travelled out to one of the forestry coops and been photographed beside the stump of a large clear-felled karri. Coming from someone of Malthouse's background this was tantamount to a declaration of war. Malthouse was not just any football coach; he had a profile as one of the toughest and most dour of the AFL priesthood, a 'real man' if ever there was one. In 1998 he had surprised the local community by declaring that the prevention of further devastation by developers in Western Australia's south-west was more important to him than his job as a football coach. In a culture where football amounts to a kind of religion this was a big statement, and a huge publicity bonus for the environment movement. 'From meanie to greenie' commented the *Age*, which ran a long article on Malthouse's views in its sports section in the year he returned to Victoria to coach Collingwood. Back in his home state Malthouse wasted no time in making plain his opposition to the destruction of Victoria's last remaining stands of box-ironbark forests. 'His engagement with the environmental issue, revealed in Perth, has grown into a form of crusade,' wrote reporter Greg Baum. The national assistant secretary of the Construction, Forestry, Mining and Energy Union's forestry division, Michael O'Connor, responded by branding Malthouse 'an idiot'. Undeterred, Malthouse continues to speak out on Kyoto, salinity and Bougainville. Baum asked him what he would like to do when he retired from football. 'I would love, when that time came, to go and drill a well, build a mud-brick house. I'd like to put something back into the world. I'll wake up one day and say, "That's it." And I'll go hug a tree.'

The message from the Western Australian state election of February 2001 was this: after a period of stagnation in the nineties the environment was back on the political agenda. The feeling that some kind of second wave was building would soon be confirmed in Queensland's state elections in April of that year. In this poll the same Greens who could manage only 2 per cent in the Senate election of 1998 averaged 7 per cent across thirty-one electorates — almost double what they had polled three years earlier and only 1 per cent less than the result in Western Australia. In the subsequent by-election for the federal seat of Ryan, campaigning on the slogan 'Never forget the Democrats' GST', the Greens again polled a healthy 6.3 per cent. According to the Convenor of the Queensland Greens, Drew Hutton, 'some members left the Greens over the GST slogan — you're not supposed to be negative or criticise people.' For the first time in Ryan the Greens had polled higher than the Democrats. Scrutineers and informal exit polls indicated that land clearing had also been a major issue in the Greens' gains.

In the wake of Western Australia's and Queensland's state elections, support for green politics continued to diversify in interesting ways. Suddenly there was a proliferation of special groups identifying themselves by vocation such as Lawyers for Forests, Doctors for Forests and Teachers for Forests. Keith Woollard from Western Australia visited Victoria to discuss election strategies with Doctors for Native Forests who, according to a report in the *Age*, had the 'tacit support' of the AMA. Liberals for Forests set up branches in Victoria and Tasmania and in a separate reaction to the loss in Western Australia, Victorian Liberal MPs announced the formation of Liberals for the Environment, claiming to have an informal network of fifty Liberal members, party activists and 'experts' in conservation. On 10 May 2001 Lawyers for Forests was launched in Victoria by the Federal Court's Justice Murray Wilcox. Its president, Lucy Turner, cited the Western Australian elections as its inspiration. 'There is absolutely no doubt that there is a renaissance going on,' said Ms Turner. 'The environment is an issue that will not go away.' The *Sunday Age*'s Environment Reporter, Claire Miller, wrote: 'It would be a mistake to say the Greens Party is the political arm of this distinctly bipartisan movement; it is too diverse a constituency to be pigeon-holed like that. It is a social movement in that different elements are spontaneously self-organising around a set of common aims, which the new green wave wants to translate into votes in Victoria and elsewhere … The significance of such amorphous movements as those formed by the doctors and lawyers is difficult to assess. There are no leaders who speak for all, no formal overarching organisation for which the pollsters can gauge support, no one with whom to strike preference deals. Nonetheless their political clout cannot be underestimated. They can come from nowhere, as in the WA election, to upset the status quo' ('Greens in suits out of the fringe, in for the forests', 1/4/2001) While the lawyers had no intention of contesting elections, there were doctors who thought otherwise. Twelve months after their formation, the new Doctors for Forests in Tasmania would contest the Tasmanian state elections.

The outcome of the Tasmanian state elections in July of 2002 is one of the great comebacks in Australian political history. To appreciate the extent of the Greens' triumph and their recovery from a seemingly impossible position it is necessary first of all to recapitulate on the cynical attempt by both parties to eliminate them in 1998. At that time the Tasmanian ALP was still clinging to a legacy of bitterness at the outcome of the Accord in 1991. At the instigation of then Leader of the Opposition (now Premier) Jim Bacon it put forward a proposal to change the electoral system in such a way that the prospect of minor parties securing representation would be all but eliminated. On a populist platform of reducing the number of politicians overall, the Labor opposition and Liberal government combined in 1998 to reduce the number of members in the lower house from thirty-five to twenty-five. Under Tasmania's unique Hare Clark system of proportional representation, this resulted in five members per electorate rather than seven. More to the point, it now meant that in order to get elected a candidate must poll approximately 17 per cent of the vote instead of 12 per cent as before. The Hare Clark system had long been regarded by international scholars in the field as the most democratic in the world and it had taken forty years to fine-tune the Tasmanian version of it. At the very least any proposed changes to it should have been put to referendum. Now, overnight, it was truncated by political fiat. In the subsequent election, Green representation was reduced from five to one.

No one, including many of their own supporters, believed that the Greens could bounce back from the lock-out of 1998 but just four years later, in July 2002, they stormed back into Tasmania's lower house with an unprecedented 18 per cent of the vote – not only their highest vote ever but, according to Brown, the highest Green vote recorded in any general election anywhere in the world. This time there was no Franklin River or Wesley Vale pulp mill. Instead, there was widespread disgust at the government's aggressive take-it-or-leave-it forestry policy. Beyond even this, however, there was an increasing reaction to the merging of both parties into an all but seamless conservatism, an ongoing move to the right in which real policy differences were negligible and phoney debates around issues of Sunday trading or whether Tasmania could afford an AFL team dominated the campaign. Within four years the one Green parliamentarian, Peg Putt, had become widely referred to in the community as 'the real opposition'. Such labels arise out of grass-roots perception and are not easily dismissed, so it might well be said that the major parties ought to have been less surprised by the Greens' comeback than they seemed to be. In any event, at the end of counting the Greens had gained four seats as compared to the Liberal Party's seven (Labor held government with fourteen). 'We are now one of the major parties,' announced Greens leader, Peg Putt, who doubled her primary vote. In her southern electorate of Denison, formerly Bob Brown's electorate, the Greens polled an astonishing 25 per cent of the vote to the Liberals' 22 per cent. And it was interesting to note the backgrounds of the other successful Green candidates. Two of the four were rural while one, Tim Morris, is the former mayor of a depressed region in Tasmania's south, an ex-farmer and ex-sawmiller who cites the Franklin River blockade as a formative influence. Kim Booth is another one-time sawmiller who claims to have been sickened by the clear-felling of Mother Cummings peak in the Meander Valley in 1998. The fourth Green, advertising consultant Nick McKim, had been arrested at the Farmhouse Creek demonstration. The interesting thing about all of these candidates is that each has a lineage of activism that goes back a number of years.

Unsurprisingly there was a strong emphasis in the media on the comfortable Labor victory – 52 per cent of the vote – and the dismal showing of the Liberal opposition, whose vote was reduced to around 27 per cent. But even apart from a damaging dogfight for the Liberal leadership, the real import of the Labor victory was the degree to which a centre-right state ALP government representing a middle-ground, business-oriented constellation of interests had successfully appropriated the traditional Liberal constituency. Anecdotal evidence abounded of Liberal voters who for the first time had voted Labor. Labor was acknowledged as the more effective and stable of the two conservative parties on offer. Meanwhile the progressive voters in both major parties defected to the Greens.

In all of this, forestry policy was a key issue. Tasmania has only 25 per cent of the forest cover that it had when the Europeans arrived. Today it exports more woodchips than all the other states and territories put together, with a 19 per cent increase since the Tasmanian Regional Forest Agreement was introduced in 2000 and a haul of statistics on clear-felling that are daunting. If Graham Richardson thought that woodchipping was a scandalous waste of resources, few in the Tasmanian wing of his party agree with him. The Greens claim that the state ALP is now as much in the pockets of the forestry industry as ever it was under the sway of the Hydro-Electric Commission, and for a much inferior cause. At least the HEC had a vision.

Ironically it was the Liberals who suffered most from the 1998 reduction in the numbers of the parliament, outwitted by the ALP. It is also ironic that when Bob Cheek first became Liberal leader he began by expressing doubts about Tasmanian forestry policy. After the election Cheek stated that he thought many of his party's policies were too right-wing and it had to return to the centre of politics (presumably to the left of the Tasmanian ALP). He also claimed that, believing there was scope within the Regional Forest Agreement to end clear-felling, he had pushed the Liberal Party to go to the election on that platform but that the threat of a party split had stopped him. In saying all this Cheek revealed the degree to which he had been the victim of his own political timidity since throughout the campaign he had colluded with Bacon to erase the forestry issue from the agenda. As journalist Simon Bevilacqua wrote, forestry was 'the phantom issue' of the election. No one in the leadership of the major parties wanted to talk about the 'f' word and some were upset by what they regarded as a tactical blunder when the woodchipping giant Gunn's Ltd ran large and expensive advertisements in the local paper arguing for 'stable government'. (In a practice worthy of George Orwell's newspeak the phrase 'stable government' had become the widely used euphemism for maintaining current forestry policy and not voting Green.) One of the directors of Gunns Ltd was none other than former Liberal premier Robin Gray, who had fought so aggressively to flood the Franklin River and challenged the Hawke government in the High Court. When it came to industry policy he was prepared to give de facto support to Labor against his own party just as former Labor premier Eric Reece had openly supported Gray against Federal Labor over the Franklin dam.

In the spirit of their Western Australian confrères there were, however, other Liberals who were prepared to admit to reconstruction. John Freeman, former lord mayor of Hobart and campaign manager of one of the successful Liberal candidates in Denison, came out during the election to voice his opposition to Tasmania's 'primitive' forestry practices.

'It might once have been a Greens issue, and it might have been amusing to poke fun at ferals, but that's not on any more,' he said. 'The state government must recognise that it is a community issue — there is as much feeling in conservative circles as anywhere — and it will not go away.' Freeman might have been just another born-again pollie hoping to regain the mayoralty on a green platform, or he may be another flag on the map of that evolving constituency, the small 'g' constituency. As for the Doctors for Forests, they were politically inexperienced and their low vote served to demonstrate that in the cut-throat play-off of the old stagers, good intentions are not enough. This is not an arena in which people can successfully 'come from nowhere'. It takes a long time to produce a disciplined and effective party organisation, as the fate of One Nation demonstrates. The Greens on the other hand are long-term and experienced campaigners. From the grass roots up they have evolved not only an effective leadership but an army of volunteer labour. Their results in Tasmania in July 2002 demonstrated not only that the canary in the mine is still whistling, but that it's chirping louder than ever.

The election I have so far overlooked is the federal election of November 2001. There is a sense in which state elections can appear on the national electoral scene as village carnivals of the national mood: all local character and regional idiosyncrasy; bizarre bush tirades and vicious backyard brawls. What is more telling over time, however, are the detailed statistics of federal elections; the psephologist's passionless graph of the zeitgeist or, as they say in the strategy forums, the big picture.

The federal election of 2001 was a sideshow alley of surpassing ugliness and the bitter feeling it engendered lingers on in a good deal of pessimism and dismay among progressive voters. But in all the fuss about *Tampa* and the more recent colour and movement of Meg versus Natasha, what has largely been overlooked are the long-term trends.

In the 2001 House of Representatives elections the Greens, along with the Democrats and the ALP, fielded candidates in every electorate. In what has been interpreted as an outcry against the Coalition government's *Tampa* policy the Greens received their greatest primary support from inner metropolitan areas which were safe seats held by either of the major parties. While this may have appeared to be wasted support (no Green candidate was elected in the House of Representatives), Greens preferences exerted an influence on the outcome of nineteen seats.

The highest Greens vote was 15.7 per cent in the seat of Melbourne, an inner metropolitan, safe Labor seat. The Greens also had good support in Sydney (14.7 per cent) and Grayndler (13.06 per cent), both inner metropolitan, safe Labor seats. In Kooyong, an inner metropolitan, safe Liberal seat, they received strong support at 10.7 per cent and in Denison, an inner metropolitan, safe Labor seat, they received 10.5 per cent of the vote. A relatively strong Greens vote also came from other inner metropolitan divisions in Brisbane (7.8 per cent), Fraser in the ACT (7.9 per cent),

Adelaide (6 per cent) and Curtin (9.2 per cent). Curtin is an inner metropolitan, safe Liberal seat in Perth.

As these figures indicate, the strongest primary vote for the Greens came from inner metropolitan areas in Melbourne and Sydney with other capital cities recording higher support than the country and outer metropolitan divisions. Most of the inner metropolitan seats referred to above are classified by the Australian Electoral Commission as safe; preferences therefore did not play a large part in determining final outcomes. For example, in Grayndler the Greens' 13.06 per cent was not decisive in giving the ALP's Anthony Albanese (who received 49 per cent) his seat. In the seat of Sydney, however, the sitting ALP member Tanya Plibersek received 44 per cent of the vote and had to rely on Greens or Democrats preferences.

Greens preferences played an important role in many other divisions in the 2001 election. In the marginal seat of Jagajaga, won by Labor's Jenny Macklin with 45.6 per cent of the vote, Macklin had to rely on Greens and Democrats preferences. This was also the situation in Melbourne Ports where the successful ALP candidate, Michael Danby, received 39.4 per cent of the vote and the Liberal Party candidate 39.7 per cent. The Greens had a vote of 11.3 per cent in that division. Preferences were also decisive in McMillan where again the ALP candidate was helped over the line by Greens and Democrats preferences. Other Victorian divisions where Green preferences played a decisive role were Chisholm, Isaacs, Bendigo and Ballarat. This adds up to seven divisions in Victoria where Greens preferences were important if not decisive in giving seats to ALP candidates.

In New South Wales the influence of Greens preferences was less than in Victoria. Notable seats were Lowe and Newcastle where Greens and Democrats preferences were decisive in giving the seat to the ALP. Even more interesting, perhaps, is the fact that for the first time the Greens polled respectably in rural areas of New South Wales. A scan of their vote in the sheep-wheat belt shows the Greens polling up to 3–4 per cent in areas hitherto impenetrable to the traditional Left.

In Queensland the marginal seat of Brisbane went to the ALP through Greens and Democrats preferences as did Bowman. In Western Australia the outer metropolitan seat of Canning went to the Liberals on preferences from One Nation with Greens and Democrats preferences contributing. In Hasluck the ALP came from behind to win on Greens and Democrats preferences, while in Stirling the successful ALP candidate had only 41.2 per cent of the vote and had to rely on preference support from a solid Greens and Democrats vote. In the division of Swan there was a similar situation.

In summary, Greens preferences played an important and at times decisive role in nineteen electoral divisions. Of those nineteen, sixteen are now held by the ALP and three by the Coalition.

The Greens vs the Democrats

On 2 August 2002 the *Australian* published a Newspoll that showed national support for the Democrats to be at an all-time low of 3 per cent while the Greens had a high of 6 per cent. The polls also indicated there had been a continual fall in support for the Democrats since the 2001 federal election and a small rise in support for the Greens over that time. In the September poll, support for the Greens rose a further 2 per cent (making it just half a per cent under the German Greens' showing in the recent elections there) with a majority of these votes coming from Labor. To anyone who has made a close study of the last three federal elections there is no way that these figures could have come as a surprise.

In relation to the House of Representatives, from 1998 to 2001 the Democrats secured an increase in support of 0.28 per cent to gain 5.41 per cent of the national vote, an increase of 2.34 percentage points. In comparison the Greens received 4.96 per cent of the national vote, an increase of 2.34 percentage points. The Democrats vote in the House of Representatives increased in line with the national trend of the last two decades that shows an increasing number of votes going to the smaller parties. The significant thing here, however, is the ongoing trend that emerges from these figures. This marginal increase by the Democrats of 0.28 per cent is significantly less than the increase in Green support which was a steady 2.34 per cent.

Overall the Greens recorded stronger support than the Democrats in those states where Green parties had been established for longer, namely Tasmania, Western Australia and New South Wales, while the Democrats out-polled the Greens in other states and territories. Not surprisingly the Democrats received their strongest support from South Australia, their home state. In total, the older party, the Democrats, had a stronger supporting vote in 83 of the House of Representatives electoral divisions, while the Greens' percentage was stronger in 67 divisions.

If the growth in Greens support continues, then on these figures alone there is every reason to expect that the Greens will become the dominant minor party at the next federal election.

The Senate

The national trends in the House of Representatives are also reflected in the last three Senate elections.

The table overleaf clearly shows that over the last three Senate elections the Democrats vote has declined in each election. From a high of over 1,179,357 or 10.82 per cent of the national tally in 1996, it has dropped to 843,000 or 7.25 per cent in 2001. In the same period the Greens vote has consistently grown. The small figure of 180,000 in 1996 had grown to 510,000 last year. If the figures for the Greens in Western Australia are added to these, we see that in 1996 the Greens in total received 237,410 votes or 2.18 per cent of the national vote. This figure grew to 574,550 or 4.94 per cent in 2001. The figure at the 2001 election places the Greens approximately 270,000 votes behind the Democrats which is pretty much the amount of increase in Greens votes from 1998 to 2001.

On these figures alone it is clear that if these trends continue to the next election the Greens will outstrip the Democrats in both the Senate and the House of Representatives.

Senate	1996 Enrolments: 11,740,568		1998 Enrolments: 12,154,050		2001 Enrolments: 12,708,837	
Party	Votes	%	Votes	%	Votes	%
LP	1,770,486	16.24	1,528,730	13.63	1,824,639	15.69
NP	312,769	2.87	208,536	1.86	222,860	1.92
LNP	2,669,377	24.49	2,452,407	21.87	2,776,089	23.88
CLP	40,050	0.37	36,063	0.32	40,680	0.35
ALP	3,940,150	36.15	4,182,963	37.31	2,691,415	23.15
ACL	–	–	–	–	1,299,488	11.18
DEM	1,179,357	10.82	947,940	8.45	842,984	7.25
GRN	**180,404**	**1.66**	**244,165**	**2.18**	**509,814**	**4.38**
HAN	–	–	1,007,439	8.99	644,346	5.54
CDP	–	–	122,516	1.09	–	–
DLP	–	–	29,893	0.27	–	–
GRN WA	**57,006**	**0.52**	**61,063**	**0.54**	**64,736**	**0.56**
HAR	–	–	24,254	0.22	–	–
UNI	–	–	93,968	0.84	–	–
OTH	749,438	6.87	271,966	2.43	710,478	6.11
FORMAL	10,899,037	96.50	11,211,903	96.69	11,627,529	96.11
INFORMAL	395,442	3.50	375,462	3.24	470,961	3.89
TOTAL	11,294,479	96.20	11,587,365	95.34	12 098,490	95.20

In New South Wales in 2001 the Greens won a Senate seat (Kerry Nettle), and over the last two Senate elections the Democrats vote has declined while the Greens vote has increased. A continuation of this trend in the next Senate elections of 2004–5 would see Democrats member Aden Ridgeway lose his seat to a Greens candidate (and this is on statistical trends alone and not taking into account the recent turmoil among the Democrat leadership).

In Victoria the 2001 Senate vote for the Greens (174,817 or 0.42 per cent of a quota) was greater than in New South Wales (169,139 or 0.31 per cent) although the preference flow had a different outcome and the Greens candidate, businessman Scott Kinnear, lost out to the Democrat Lyn Allison. An important element in the next Senate election in Victoria is that there is no sitting Democrats member coming up for re-election in 2004–5. The field will therefore be wide open in regard to the minor parties. The growth in Greens support (70,800 in 1998, 174,800 in 2001) is an increase of 100,000 votes and represents an increase of well over 100 per cent in their total vote (from 0.17 of a quota to 0.42 per cent of a quota). There can be no doubt that feeling over major party policy on the *Tampa* crisis contributed greatly to this jump but the direction of the increase is consistent with longer-term trends before and after 2001. Given the Victorian surge in support for the Greens since 1998 we can expect growth to continue so that after the 2004–5 Senate elections, Victoria will have a Greens Senator.

In Western Australia the Greens and Democrats have been neck and neck over the past two Senate elections. The Democrats vote has declined from 68,100 to 64,800 while the Greens vote has increased from 61,100 to 64,700. Again, on current trends, the sitting Democrats member Brian Greig will lose his seat to a Greens candidate. Since Greig is a long-time political operator one can assume he knows how to read the trends and this may have prompted him to make a bid for the Democrat leadership, and with it a higher profile in the electorate.

In Queensland, numbers for the Greens are still relatively small but the increase is healthy while here also the Democrats vote is declining. In the 1998 Senate elections the Democrats received 156,450 votes which then declined in 2001 to 143,900 votes. This decline contrasts with the increase in Greens votes from 42,260 in 1998 to 71,100 in 2001. Queensland has a sitting Democrats member, John Cherry, whose term expires in 2004–5. Based upon the voting trends of each of these parties the Democrats would return a senator in 2004–5. Such a conclusion, however, does not take into account any increase in the long-term drift away from the Democrats occasioned by their recent troubles.

In Tasmania the Greens are now clearly the dominant minor party, winning over the Democrats in all of the five divisions in the House of

Representatives and also in the Senate. In Tasmania the Greens also have their healthiest Senate vote in Australia with 0.97 per cent of a quota, or 42,570 votes. In 1996 Bob Brown first won a seat – with 26,800 votes, or 0.61 of a quota – over sitting Democrats member, Robert Bell, who, despite having solid support of 22,000 votes or 0.50 of a quota, lost to those small 'g' green voters who made a strategic switch. The Democrats vote then dropped to 12,100 in next Senate election of 1998. The 2001 election figures strongly suggest that the Greens have now mostly taken over the Tasmanian Democrats vote. At the next Senate election there is no sitting Democrats or Greens member coming up for election. There may, however, be the perennial independent Brian Harradine, elected to the Senate in 1983. Given his age and that his vote has been steadily declining since then, Harradine may choose this moment to retire. If the Greens endorse the widely known and highly respected former leader of the Tasmanian Greens parliamentary party, Christine Milne, there is a better than good chance that the Greens vote of around 40,000 will hold and Milne will be elected. Like Harradine, Milne has strong ties with the Catholic Church and its constituency so that even if Harradine re-contests, it's possible that Milne could defeat him.

In South Australia the Greens appear less likely to make significant gains over the Democrats in the short term. The Democrats have stronger voter support in each one of the twelve House of Representatives divisions. In addition, the lead the Democrats have over the Greens in the Senate vote is also greater in South Australia than any other state. In 1989 they received 117,620 votes or 0.87 per cent of a quota while in 2001 they had a slight increase in support of 122,200 votes, or 0.88 of a quota. In comparison, over the same elections the Greens received 20,890 or 0.15 per cent of a quota and in 2001, 33,440 votes or 0.24 per cent of a quota.

The factor that could make the 2004–5 Senate election in South Australia less predictable for the Democrats is the former leader Meg Lees. Her term expires in 2004–5 and she may want to be a candidate again. By then she may have returned to the Democrats fold but on present indications this is unlikely. If she stands as an independent then leakage in the Democrats vote is likely to be considerable.

The ACT and Northern Territory elect only two senators each election. This means that the minor parties have no chance of winning a seat until their vote increases to around 33 per cent of the vote.

The Greens' triumph in the Cunningham by-election of October 2002 is in some part due to special factors, among them the decision by the Liberal Party not to field a candidate. Nevertheless a range of opinion polls suggest that a momentum is building and the Greens are set to play a significant role in the coming state elections in New South Wales and Victoria. In one way it's easy to be wise after the event about the Greens' new electoral standing, and to point to any number of local influences (the Sandon Point dispute in Cunningham) or new foreign policy anxieties (anti-war feeling on Iraq) that contribute to it. My point here, however, is that if we look at election results in the last three federal elections, and block out everything since, a clear underlying trend is revealed in the figures, a trend that highlights the emergence of that new ecological constituency which I have argued for elsewhere in this essay.

The *Tampa* crisis may have significantly enlarged that tendency but, as the table on page 60 makes clear, it did not create it. Nor is the factor of Bob Brown's election to the Senate in 1996 to be underestimated as a factor in the 2001 Greens vote. With Brown's presence in the Senate the Greens Party acquired, for the first time, an impressive and outspoken leader at the national parliamentary level.

Politics is notoriously hard to predict and only a year ago the Democrats were being led by Natasha Stott Despoja as one-half of a so-called 'dream team'. With the open fracture of the Democrats and the so-called 'bleeding from the Left' of the Labor Party, not to mention disaffected small 'l' Liberals, it now seems even more likely that we will see a continuation of the long-term trend towards the Greens. The problem for the Democrats never was Natasha Stott Despoja's leadership or putative radicalism. The problem lies in the way the Democrats have handled their success, i.e. by taking up their more conservative options, thereby becoming less attractive to those increasingly disaffected elements of the electorate who are casting about for stronger moral leadership on peace, social justice and the environment. Even so, long-time activist and Convenor of the Queensland Greens, Drew Hutton, is cautious in his reading of the trends. 'I'm not super-confident of building a constituency beyond the 5–7 per cent. As I said … a lot of voters are strategic and they'll vote for what they think is going to work at any given time. We've got a core constituency of probably 3 per cent. But that could and will go on for quite a long time. This is the imperative of the twenty-first century, the ecological imperative. The material base of life in the twenty-first century is a deep reason for Green electoral growth and it has nothing to do with the Democrats as such. The problem is not whether Natasha Stott Despoja is the right leader, the fights in the Democrats are a symptom of their problem, not its cause. And a lot of the people who came over to the Greens on refugees haven't gone back. I know a lot of them and they stop me in the street and tell me, we're not going to do anything else but vote Green now. And what they're doing is recognising that for twenty years Labor has been heading this way. Labor's core vote is now only 30 per cent. Voters hung in and hung in and *Tampa* was the last straw.'

However we choose to interpret the 2001 figures one thing is clear: over the last two decades one of the key electoral trends in this country has been a decline in support for the major parties. From an average of 92 per cent in the eighties it fell to 84 per cent in the nineties, 79.6 per cent in 1998 and approximately the same in the 2001 election. In the New Zealand election of July 2002 the major party vote fell precipitously to around 63 per cent. In the view of many analysts these figures are an indication of declining levels of trust in the major parties. 'I think there's been a bit of a flip-over', says Peter Hay, 'from what I would describe as a healthy cynicism towards politicians to a deep-seated alienation from the political process as such.' Hay, author of *Main Currents in Western Environmental Thought*, is Director of the University of Tasmania's Environmental Studies Centre. He is also a former activist in the Labor Party. Like many others he ascribes the increasing volatility of the electorate to the adoption by both parties of the ideology of economic rationalism. 'No longer do we have class-based parties of capital and labour: since the time of the Hawke/Keating deregulation what we have instead is two parties of capital. This means an intelligent, questing middle-class vote which now has no political home.' Despite this, Hay is reluctant to endorse any case for a paradigm shift. For him the Green vote, at least at the federal level, is still largely a soft vote that the Labor Party can recover by 'tooling some policy changes'. In his view it takes at least two and probably three trips to the ballot box with no change before a voting 'habit' comes to constitute a semi-permanent allegiance. (This does not, he emphasises, apply to Tasmania where the Green vote is 'locked away' at 9–10 per cent.) On the other hand, 'If the Labor Party doesn't change course, does not make any effort and say, "Well, we've lost this vote and how do we get it back?" then over a period of time it will become an ossified party.' When I ask Hay where this policy shift in the ALP is going to come from – where are the sources of regeneration within Labor? – he shakes his head and says

that at present he can't see any. Hay himself is no longer a member of the
Labor Party.

It is often difficult to remember what the Australia political culture was
like in the seventies. The term 'ecology' had little if any public currency.
Conservation was something we did to old buildings and the Department
of the Environment issued licences to industry and local government to
pollute. In other words, there was no Green constituency.

At the end of the sixties the central focus of the Labor Party was still
its well-defined constituency of employees, both blue- and white-collar
workers. In those days the Labor Party believed in social values as the
political bottom line. The Coalition's constituency was less well defined
but it nonetheless had an established base in the business community
together with a country constituency plus the forgotten people, as Bob
Menzies liked to call that section of the middle and lower-middle classes
that voted Liberal. These were reasonably well-defined constituencies with
established interests that were more-or-less represented by the Labor and
Coalition parties. This situation began to change, slowly at first and then
in the last two decades with gathering momentum.

From the early fifties the Liberals used an effective strategy to undermine Labor's representation of employees. In essence it consisted of
attacking the unions, painting them as socially irresponsible and then pro-
jecting that negative picture onto the Labor Party. This was easily done by
pointing to the structural role of the unions within the Labor Party and by
invoking the legendary thirty-six faceless men. The point of this strategy,
which can still be effective today in arguments about the level of union
representation within the ALP, is to de-legitimise and render inauthentic
the traditional bedrock of the Labor Party's representation; to weaken the
links between the constituency and its parliamentary representatives.

We could argue about how effective this strategy was in keeping Labor
out of office federally before 1972. There is no doubt, however, that from
Whitlam onwards the strategy of attacking the ALP through the unions

was superseded by a concerted campaign conducted outside of parliament. In his book *Taking the Risk out of Democracy*, Alex Carey records how the election of a Labor government in 1972 'shocked Australian capitalism greatly' and he traces in fine detail the business-led campaigns of the seventies and eighties that set out to eliminate the possibility of any further Whitlam-style 'risks'. Not the least effect of these campaigns, wrote Carey, 'has been to generate public cynicism in the capacity of governments to protect, represent and enhance the public interest'. These campaigns have been underway for the best part of thirty years now but remain largely ignored by both the media and the academy. Carey estimates that between $8 and $10 million per year was spent in the eighties on promoting 'economic education' in most educational institutions throughout the country, and in propagating the New Right agenda of economic rationalism through 'think tanks', public relations and the media. (Some of the more notable examples are the Enterprise Australia Association with its Schools and Colleges Programme and the Advance Australia Campaign which I have documented elsewhere.) The aim of these operations was to promote a business ethic and a politics of the financial bottom line, and their chief outcome has been to create a large middle-ground constituency manufactured out of the tacit consent of millions and facilitated by well-oiled public relations campaigns that position everyone as a consumer and a shareholder. Across this amorphous middle ground it is easy to blur the line between the social and the financial in favour of the latter, which is then posited as the only possible means to the former (remember the 'trickle down effect'?).

Today there is only one major constituency but one that is equally well represented by either of the major parties. As the de-legitimisation of public ownership and community values has advanced, the Labor Party has moved further to the right in its desire to occupy the central ground of this manufactured constituency. In the eighties when the business-led campaign was in full swing, the Hawke government sought to distance itself from the social democratic reforms of the Whitlam government and began to highlight 'the economy' as the first principle in responsible government. Like the Liberals it took on and promoted the business agenda without embarrassment or inhibition. Issues that in the past were traditional Labor concerns, like a liberal education, were downgraded. From being seen as having a value in its own right, education became an instrument, a way to serve the aims of industry.

Labor and business now share the same view of things in which the bottom line is economic, not social. They may differ in emphasis but they share the same paradigm. Meanwhile the ALP continues to live in fear of the ghosts of its past. When the *Tampa* crisis arose, at least some of its leaders must have shuddered at the recollection of taking a moral stand on Vietnam in the election of 1966, and how the party had been buried in an electoral landslide. Because cultural memory is selective, no one reminded the electorate of the cynical manoeuvres of the Coalition in the past and of the great price that Labor under Calwell had paid for its honourable defeat. But Labor remembered, and gambled on the 'me-too' option to achieve either a narrow victory or a loss with minimal damage. This does not, however, explain why, having lost the election and changed leaders, the ALP continues to regard economics as its ground zero and to articulate the jaded and dispiriting rhetoric of the conservative centre.

The reaction to all this by the progressive supporters of both major parties has been a steady drift into a new and growing constituency by default. This leakage to smaller parties and independents will continue to increase as the new progressive constituency demands, among other things, a greater social and ecological accounting. Part of this new constituency is a broad-based protest vote and part is a core ecology vote, and no one at present can be sure of the relative proportions of each, though the European experience suggests that the core vote is enlarging at a rapid rate, accelerated by pollution scandals, a distrust of bio-technology and, above all, by a perception of global warming.

In Australia it is the Greens, not the Democrats, who have emerged as the authentic representatives of this developing constituency and I call them authentic for a number of reasons. Firstly, theirs is an organic leadership that has evolved out of prolonged grass-roots activism. They are not a collection of ersatz Liberals. Secondly, the Greens are clear on their bottom-line accounting where the Democrats are not. There is a crucial sense in which the Greens know where they come from. Already they have a sense of tradition, their own potent history and folklore, whereas the Democrats tend to re-invent themselves with each new change in leadership. Thirdly, and of growing significance, the Greens belong to an international movement that is increasingly co-ordinated in its organisation and which underscores the ecological bottom line. By contrast, and like the Democratic Labor Party (DLP) before them, the Democrats do not have a core constituency to represent. They are a barnacle on the bow of the major party constituency. Because they do not have a core constituency they do not have a political accounting that is particular to them. Sometimes they accept a bottom line that is financial (the GST), sometimes they accept a social bottom line and sometimes an environmental one. What they mainly represent is the notion of a general brake on government (keeping the bastards honest) and this is a reactive value that will not be sufficient to sustain them. When the New South Wales president of the Democrats, Cameron Andrews, points to the increased success of the Liberal Democrats in Britain as a model for the Australian Democrats, he does not take into account that the UK Democrats grow out of a long tradition of British liberalism that has had parliamentary representation for well over a hundred years. The Australian Democrats have no such anchorage. In a tacit acknowledgement of this, one of the founders of the Democrats, former senator Michael Macklin, has urged them to amalgamate with the Greens.

Unlike some others, Macklin does not subscribe to the many rationalisations offered for the recent electoral gains made by the Greens. These tend to follow the line that the Western Australian election was a one-issue wonder, soon to be annulled by the implementation of the state ALP's policy on the anti-logging of old growth forests; the Tasmanian and New Zealand elections were the outcome of weak or fragmented conservative parties; and the 2001 federal election was a reaction to the *Tampa* crisis. Others eschew such rationalisations. To them it is clear that a new constituency is developing even if they don't like it. To speak off-the-record with apparatchiks from both Liberal and Labor camps is to be told in no uncertain terms where the real enemy is perceived to lie. 'The Greens are our biggest nightmare,' to quote one ministerial staffer in the federal sphere, and this may account for the level of vitriol directed towards them not only now but over the past two decades. Current appraisal of the Democrats and their woes largely adopts a tone of patronising sympathy, and the rhetoric of their adversaries and in newspaper editorials was never harsh. Though they were accused of being pixilated (the 'fairies at the bottom of the garden' jibe) they were acknowledged on the whole as good, respectable folk who, understandably, had become disillusioned with the major parties. No one has ever really feared the Democrats but the level of invective against the Greens reveals another story. Hugh Morgan of Western Mining calls them the new enemy of capitalism, replacing the old enemy of socialism. Peter Walsh agrees. The former Labor senator and finance minister in the Hawke government wrote an opinion piece in the *Australian* newspaper recently (6/8/02) in which he warned of a crisis that 'could threaten our high living standards' should the Greens' influence and support continue to grow around the country. Walsh went on to describe the Greens as 'dangerous' purveyors of 'ignorance and delusion'; 'strident' and 'absolutist' agents of 'extremism'; saboteurs, fanatics and ruthless blackmailers. Before the New Zealand election voters were warned in a series of editorials in the *New Zealand Herald* that 'prominent citizens' (read Sam Neill) 'should not endorse the Greens' electoral stunt' (read extending the GM moratorium). Employing the preferred euphemism of a decade of anti-Green press in Tasmania, the *Herald* also cautioned voters that the Greens would only jeopardise 'government stability'. After the Cunningham by-election the *Australian* (23/10/02) ran a vehement editorial accusing the Greens of being anti-everything.

Behind these choleric attacks is a strongly held suspicion that the Greens, and the constituency they represent, are a real and growing threat to the status quo. New Zealand's Helen Clark reserves her most withering scorn for the NZ Greens while choosing her coalition partners from right-wing fundamentalists who oppose much of what she has spent a lifetime campaigning for. Among Australian Laborites it is not uncommon to encounter an almost irrational resentment of the Greens as interlopers who have somehow blundered onto the political scene and stolen the ALP's moral thunder. Much of this feeling derives from an uneasy sense that in taking on board the economic agenda of the Right the federal ALP is in danger of losing – or may indeed already have lost – its traditional status as the authentic voice of social justice. 'In an era of political opportunism and unabashed pragmatism,' writes Hugh Mackay in a recent opinion piece, the Greens appear to be offering a new 'moral compass ... they are the closest thing we have to a political party that proceeds from a clear sense of its own meaning and purpose.'

This clear sense of purpose is something that will come under increasing pressure as the Greens enlarge in both membership and electoral support. Press coverage of the recent Green victory in the Cunningham by-election has focussed almost entirely on the current woes of the Labor Party. Little or no analysis has been made of the reasons why the Greens were well placed to take the seat when the fed-up-with-Labor protest vote might so easily have gone to the disaffected union nominee and independent candidate, Peter Wilson. I, for one, thought he would do better. But what is still underestimated is the degree to which Greens activists all over Australia have for a long time now been doing the hard yards at the grass roots and local government level. Sooner or later this parochial work is bound to be reflected in the federal sphere. Since the 2001 federal election the Greens Party membership has doubled across every state and now stands at around four and a half thousand and rising. According to Drew Hutton, applications for membership have jumped from one a week to an average of five since the result in Cunningham. The financial problems that have bedevilled earlier campaigns will soon be alleviated in those states where the Greens vote has reached or appears soon to reach the magic figure of 4 per cent that entitles the party to an allocation from public funding in election campaigns. In Victoria, for example, where Scott Kinnear so narrowly missed out on a Senate seat in 2001, Greens campaign funds will jump from $80,000 at the last federal election to $500,000 at the next. In addition, some unions are now making contributions to campaign funds at the state level, another indicator of the success of the Greens' project of making environmental protection a part of the national consensus, and a further indicator, too, of the ways in which the Greens continue to outflank the Democrats in the growth of the new ecological constituency. (It should also be mentioned that several unions are openly hostile to the Greens.)

Out of this very consolidation, problems will arise. Christine Milne has expressed concern that the Greens should not make the mistake of the Greens Party in the UK where celebrities with no track record have been drafted into standing on a Green ticket. Drew Hutton anticipates problems with maintaining a balance between the social justice and the deep ecology formations within the party. 'What worries me most in the Greens', says Hutton, 'are those people coming from what I would see as an ideological social justice position, ideological leftism. That traditional leftism has run its course – there's no longer any currency in it. And all that hard-headedness and practicality and intellectual honesty and focus that caused the Greens to get off the ground in the first place against all the odds could get lost in ideological fervour, and that's the biggest danger. Everyone's focussed on opportunistic politicians who get in for their own benefit, careerists, but there's a problem at the other end as well.' Some in the Democrats' camp are also alive to this possibility, and in a recent opinion piece in the *Sydney Morning Herald* Web Diary, Cameron Andrews attacked the Greens on the basis of their ties to the 'socialist left', labelling NSW Green senator Kerry Nettle (who replaced his former boss, Vicki Bourne) as a 'red'.

Some of this was thrown into relief recently when Bob Brown caused a ripple on the pond of Greens unanimity by thinking aloud in public on the issue of Telstra, asking whether the Greens Party might not consider a trade-off with the Howard government in which it supported a full sell-off in return for forest protection. When I raised this with Drew Hutton he defended Brown's kite-flying by pointing to the fact that Brown had raised the issue in the week before a meeting of the Greens Party's national council and in so doing was making a statement of his intention to raise the possibility of a trade-off of Telstra in favour of forests. 'He wasn't saying "I'm going to do it." He said "I'm going to ask the Green council." There were Greens who didn't favour the idea but they understood his feeling. It came purely out of desperation, watching the forests go.' In the event the council, which is made up of delegates elected from state conferences, voted to retain its policy of opposing any further sell-off. This moment of public tension between the environment and the social justice strands of the Greens Party reminded me of a moment in the experience of the Green–Labor Accord in Tasmania as recalled by the Tasmanian writer Cassandra Pybus. During a tense stalemate of Accord negotiations over logging practices in a disputed area, one Green strategist is said to have complained that 'a whole social agenda was going down the gurgler because of a piece of scrub'. I suspect that this comment may be apocryphal but like all apocrypha it speaks to certain ambivalences and strains. On the other hand there were many who predicted that the German Greens would be destroyed by the clash of their two major factions, the so-called 'realists' and the 'fundamentalists'. Instead they have managed to re group into a formation that goes from strength to strength.

The green leadership

Though the Greens have been an organised political party for just over a decade, as a movement they have been around for almost thirty years. In that time, and over many grass-roots campaigns, they have developed a depth of experienced leadership that seems to me currently to be under-estimated, partly because it has been much less visible than the Democrats' Senate representation in terms of numbers, although this is likely to change after the next federal election. Natasha Stott Despoja has many attractive qualities but, for all that, it is inconceivable that someone so young and relatively inexperienced could ever be elevated to the leadership of the Greens. There is a sense in which the Greens are an army many of whose members have – quite literally – put their bodies on the line and it is expected that their generals will have earned their stripes at the level of grass-roots activism. They are only their foot-soldiers writ large. In addition, in background and temperament they bear little resemblance to the orthodox apparatchiks who occupy so much space on the front benches of the major parties, and this gives them a quality of authenticity that cannot be constructed out of even the best media coaching. Nor, it must be said, is this quality a monopoly of the progressive wing of politics. Part of Pauline Hanson's appeal was a recognition among her constituency of some quality of authentic being that it is impossible to fake, a quality that makes her well-nigh irreplaceable within One Nation by any current or former apparatchiks such as David Oldfield. While it may seem an odd comparison to make, on the other side of the political spectrum that same quality is there in the leadership of Bob Brown.

Traditionally Green parties around the world have been concerned to discourage cults of personality; nevertheless it must be acknowledged that Brown has grown to occupy a singular place in the Australian political landscape. In recent years derogatory epithets from the intellectual Right have involved much ado about the so-called chattering classes and the latté set. In this way is the progressive element of the Labor Party reduced to a set of lifestyle gestures. The same ploy has been tried out on the Greens, most recently when the former Tasmanian Liberal premier Robin Gray, now a director of the woodchipping giant Gunns Ltd, insinuated on the eve of the Tasmanian state election that voting Green was just a lifestyle choice. Of course, in the most profound sense, he's right.

But the jibe of the chattering classes will scarcely wash when it comes to Brown, nor for that matter to Christine Milne and Drew Hutton. All three come from conservative rural backgrounds. Hutton's father was a country butcher and former drover, Milne's a small dairy farmer and Brown's a country policeman. As a young man Brown, like his mentor, Dick Jones, voted for the Country Party.

That Brown should now be the country's most outspoken political leader against the war on Iraq is consistent with his youthful focus as a peace activist. An early indication of his mettle came in 1976 when he conducted a prolonged fast on the chilly slopes of Hobart's Mt Wellington in protest at the arrival of the USS *Enterprise* with its cargo of 100 nuclear warheads. Although it was November the weather was bad, with cold southeasterly rain almost every day and snowfalls on the higher slopes. For the eight days of the *Enterprise*'s visit, Brown took nothing but water and came down only after the ship had sailed. During that time, several dozen of the ship's crew visited him, some to talk frankly and separate themselves from the actions of their government. And it was around this time that Brown came out as gay, in a climate a good deal less receptive, and legally protective, than it is now.

In those early years Brown lived the life almost of an ascetic, giving up his income as a GP to work full-time for the environment movement. During the long and arduous Franklin River campaign he relied on friends and fellow activists to feed him, give him a bed and lend him their cars. One of Brown's chroniclers from this time wrote of his impressive equilibrium under stress, a notable absence of either self-importance or false modesty and 'an ability to accept the necessities of life from others with neither guilt nor arrogance'.

In the many green campaigns since, Brown's unflinching persona has become an emblem of political integrity. No one, as journalist Mike Seccombe once observed, is ever in any doubt about what Bob Brown stands for. He is a noted plain speaker, and if you listen to his public statements there is a striking absence of stock political rhetoric. And this is but one area in which he projects a manifest ordinariness. Even his name is suggestive of the prosaic and commonplace, and as he has aged (he is fifty-seven) he has developed a quality of gravitas that is the opposite of charisma, that patina of glamour which is now, in the political arena, largely distrusted. Since the days of the Franklin blockade that first made him a national figure he has been bashed and shot at, mocked and derided and yet there is a toughness there that none of his political opponents underestimate, 'a sharp strategic mind that can focus with ruthless determination on a problem'. This, in sum, is the character of Brown's authenticity; that he has been dogged, honest, consistent and brave, but also that he is nobody's fool. In his conservative dark suit and at times dour public demeanour he somehow manages to suggest that the radical and the respectable can co-exist within the one moral framework, and that what we have been educated to think of as contradictory may not in fact be so. Whenever I hear him speak I am reminded of the sister of the writer Alan Marshall, Elsie McConnell, who, when asked at the age of ninety-four what she believed in, replied tersely: 'Bush spirituality and bush scepticism.'

Brown is not, of course, the only one of his type. One can think of two other radical 'ascetics' thrown up by the seventies, namely Ralph Nader and former governor of California Jerry Brown. All three share similarities though also there are crucial differences. All three made an early repudiation of the wasteful consumerism of postwar affluence and have led personal lives marked by what can only be described as minimalist lifestyles, living for long periods without stable homes, mortgages, partners, offspring or even a car. But whereas Nader was drafted into the US Green Party when it needed a high-profile candidate for the presidential election of 2000 (and went for his first wilderness walk in his late fifties), Bob Brown is an authentically organic leader thrown up by the grass roots of the ecology movement. This gives him a credibility and a moral authority difficult to impugn, and so it should not have come as a surprise when recently the secretary-general of Germany's *Die Grunen*, Reinhard Butikofer, offered his own comment on the leadership of the Australian Greens. The enormous popularity in the polls of the German Green leader, Joschka Fischer, demonstrated that the Green movement could 'profitably abandon its traditional strategy of downplaying leaders', said Butikofer. 'We found that if the person is trusted and seen as a person of principle, then people who would not normally be Green supporters are prepared to listen to the message and the policies. Bob Brown is that sort of person.'

These comments remind us once again of the degree to which the Australian Greens are part of an international movement, a significant factor in their confidence and stamina at the local level. There are now Green parties and movements in over seventy countries with thousands of Green representatives spread across all levels of government. In Europe, especially, where there is a federation of Green parties, the Greens are refining the art of coalition politics. They have several thousand local councillors, almost one thousand regional representatives and forty-eight members of the European Parliament where the Green Group is the fourth largest group. Currently they participate in close to one-third of European Union governments: Finland, Germany, Belgium and Italy (and are also a coalition partner in Sweden). In the last Belgian general election the Greens polled 25 per cent. In the previous German government Greens ministers held the key portfolios governing foreign affairs, health and agriculture (in the new government they are tipped to hold a fourth ministry). In recent times they have held two ministries responsible for agriculture within the European Union, a fact not unconnected to the strong European resistance to genetic engineering of foodstuffs and a corresponding growth in European markets for so-called clean, green produce from countries like Australia. In the US three million people voted for Ralph Nader and the Greens are the third largest party in the union and the fastest growing. In Mexico the Mexican Greens formed part of the winning coalition that overthrew a 71-year-old regime in 2001 while in 2002 Green parties fielded candidates in elections in Colombia, Korea, Papua New Guinea, New Zealand and the USA as well as in France and Germany. In Japan there were 138 local Greens politicians elected as of July 2000 (and Japanese Greens activists continue to demonstrate in Japan against woodchipping in Tasmania). In South Korea the Korean Federation for Environmental Movement (KFEM) has 80,000 members. In a well-publicised victory, it succeeded in stopping the central government from building a dam in the scenic Dong River of Kangwon province, claiming that more than 80 per cent of Koreans supported its opposition to the dam. On the minus side there are countries where it might be expected that the Green movement would be strong and yet it has failed to make an impact. Canada is one such notable case and leading environmentalist David Suzuki has been critical of the Canadian Greens for focussing too narrowly on animal rights and issues of species extinction while failing to develop a broader platform.

On the international plane the Greens see themselves as belonging to a great liberal tradition, part of a long continuum that has fought slavery, colonialism and neo-colonialism and which now needs urgently to develop a co-ordinated strategy to deal with the worst outcomes of globalisation, not least the link between poverty and sustainable development. They cite *Our Common Future*, the Brundtland Report (1987) produced by the World Commission on Environment and Development, one of whose main findings is that poverty alleviation and sustainable development go hand in hand. In April 2001 the first Global Greens conference was held in Canberra and attracted 800 people from around the world. The Greens, claims Christine Milne, are now the only grass-roots global political force on earth that is united in its philosophical world view and capable of global reach in its policy implementation.

The growth of this international ecology movement has been accompanied by the rise of an earth-based spirituality that in Western countries has gained steadily in momentum – both within and without the established churches – since the cultural revolutions of the late sixties and seventies. In the past year it has been given a boost by the appointment of Rowan Williams as the new Archbishop of Canterbury. Williams, a radical on theological as well as political issues, was arrested in 1985 for breaking into a US airbase in south-east England during a Campaign for Nuclear Disarmament protest and has been critical of Blair's first term of government in the UK. Even more interesting, perhaps, is a recent call by Pope John Paul II for a world-wide 'ecological conversion'. Humanity has 'disappointed divine expectations', declared the Pope, 'humiliating ... that "flower garden" that is the Earth, our home' and as a consequence it has become necessary 'to stimulate and sustain ecological conversion'. In response, the local organisation of Catholic Earthcare Australia (CEA), the advisory body on environmental justice to Australia's Catholic bishops, has developed and disseminated a set of educational materials for parishes, schools and community groups. This is exactly the kind of development that the Greens read as a further sign of a developing national consensus on the ecological bottom line, and it was with great fascination that I sat through a recent viewing of the CEA's video, 'The Garden Planet'. After a somewhat bland opening that features a class of primary-school children planting trees, the scene changes to reveal Tasmania's Archbishop Doyle amid the clear-felling carnage of Tasmania's Styx Valley where just 10 per cent of original old growth forest remains. The next sequence focuses on the air and ground pollution of Port Kembla – and the ominous image of a state school built right beneath the chimney stack of the copper smelter – while a voice-over informs us that every year toxic emissions of tonnes of sulphur trioxide, sulphur dioxide, lead, cadmium and arsenic are classified as legally permissible. Meanwhile health statistics in the residential surrounds are bent well out of average and children are routinely tested for their lead levels.

Contrary to the jibe that green politics is a middle class issue, new research shows that those who are most vulnerable to the effects of pollution are also those who are disadvantaged financially. Although environmental risks increasingly have a global character – nuclear fission, radioactive waste, global warming – they are always felt locally. People living near a toxic waste incinerator run more risk of contracting cancer, for example. Even global warming is felt locally and distributed unequally. A small climatic change will have great consequences for regions that are already vulnerable, such as Bangladesh, or for regions that are dependent on tropical agriculture. Risks can also be displaced by 'distributing them away' to groups which lack the social power to successfully oppose policy decisions. Research shows that a disproportionately large share of environmental risk is borne by the poor and by ethnic minorities.

German sociologist Ulrich Beck has coined the term 'risk society' to characterise 'an epoch in which the dark sides of progress increasingly come to dominate social debate'. This debate is fuelled by a declining public trust in governments and experts who have not been able to diminish the toxic outcomes associated with technology, industry or development and have often tried to cover them up, sometimes through a deliberate process of marginalising the risks, i.e. passing them off onto low-income and low-status groups. (Films like *The China Syndrome*, *Erin Brockovich* and *A Civil Suit* speak to this anxiety.) As most risks are largely invisible and it is hard to establish the cause of many environmental harms, the public relies on expert knowledge in order to identify them. This reinforces the process of marginalisation, as inequalities of knowledge and power are often intertwined.

It's not surprising then that materials on sustainable development should come from the social justice wing of the Catholic Church. These local manifestations are yet another step towards the national consensus that the Greens see themselves as instrumental in building, a consensus that will gain in support as the practices of multinationals like Monsanto are subject to further interrogation. As the New Zealand election campaign indicated, the question of genetic engineering is an electoral time bomb, and for weeks a detailed pre-election coverage of the pros and cons of GE was spread across the front pages of the *New Zealand Herald*. We in Australia are yet to have that debate in anything like so public a national forum, though it is surely only a matter of time.

No matter what our position is within the political spectrum we are most of us programmed to accept some plundering of nature as the price of progress, and the debate hitherto has turned on the question of how much. Genetic engineering of foodstuffs, however, is an issue of a different order, belonging to the same special category as nuclear weapons in that the potential damage is neither containable nor reclaimable. Thus it is that an uneasy sense grows that genetic engineering is not about the destruction of nature, which is one thing, but about its perversion, which is another thing altogether – some insupportable violation of the 'natural', of the very ground of being.

But this is not, ipso facto, a position of anti-science. The social ecologist George Myerson argues that 'ecology, the new understanding of man and nature, can become the newly legitimate face of mainstream modernity.' Myerson identifies a new role for science within a new grand narrative of ecology, a unifying vision which he describes as 'radical modernism'. Global climate change is the key to its emergence: 'With global warming science regains a credible narrative.' Myerson argues that only reliable scientific information can explain and measure the fearful effects that industrial society has produced, and that this is so even though science produced industrial society in the first place. Only reliable scientific information can forecast the weather, measure the permafrost and the glaciers and the levels of carbon dioxide. Only publicly available information that makes transparent the decision-making processes of society can provide the legitimate references for appropriate political responses. If ecology is the new understanding of our relationship to nature then this understanding arises from a free flow of information generated and legitimised by scientific method. Within the new modernity, ecology becomes a new metalanguage and a common measure of everything.

The engineers of Tasmania's Hydro-Electric Commission and the Snowy Mountains Scheme saw themselves quite self-consciously as the modern heirs of the Enlightenment and the early environmental activists as reactionaries; modern-day Luddites. Myerson's argument suggests the contrary; that it is the Greens who are the true heirs of the Enlightenment, for what they are about is the most rational management of resources, that is to say, a sustainable management of resources of the kind currently espoused by those new-wave industrial ecologists in the field of so-called 'natural capitalism'. If Myerson is right, and I believe he is, then in thirty years the ecology movement has moved from the political and philosophical fringe to a position of centrality within a revised Enlightenment project, world-wide, and the political changes consequent on this move are likely to be profound.

The greens are the bearers of this new gospel and the Green parties are the authentic political representatives of the ecological constituency: they have the history, the passion and the unifying vision. The constituency problem facing the Australian Greens Party will not lie in having too narrow a focus, or being marginal or fringe as adversaries like Helen Clark suggest. Their chief difficulty in the future will be in keeping up with the complexity of expansion within the ecological constituency and in maintaining a balance of forces within their own movement.

SOURCES

Essay sources and occasional supplementary material are given below. Page numbers indicate where the quotes, etc. appear.

ix   Lionel Trilling, *Sincerity and Authenticity*, The Charles Eliot Norton Lectures, 1969–1970, Cambridge, Mass., Harvard University Press, 1972, p.1.

2–3   Robert Hill quote: Drew Hutton and Libby Connors, *A History of the Australian Environment Movement*, Cambridge, Cambridge University Press, 1999, p.264. Transcript of a speech to the Committee for the Economic Development of Australia, 15 May 1997.

4–5   Peter Hay, *Main Currents in Western Environmental Thought*, Sydney, UNSW Press, 2002, p.2.

5   John Rodman quote: Hay, *Main Currents*, p.32.

6   On next industrial revolution: Paul Hawken, Amory B. Lovins and L. Hunter Lovins, *Natural Capitalism: The Next Industrial Revolution*, London, Earthscan Publications, 1999.

7   Smog not democratically distributed: Berenice Bovenkerk, 'Is smog democratic?', paper delivered to the *Environment, Community and Culture Conference*, University of Queensland, St Lucia, 3–5 July 2002.

7   Timothy Doyle, *Green Power: The Environment Movement in Australia*, Sydney, UNSW Press, 2000, p.19. Doyle focuses on the Queensland wet tropics campaign as a fascinating case study of how networks combine and operate over time. See p.17 for a diagram of the twenty-five organisations involved in the wet tropics campaign.

9   Max Angus: submission to the 1973 Australian Government Committee of Enquiry into the Fate of Lake Pedder.

12   Eric Reece on Lake Pedder: William J. Lines, *Taming of the Great South Land: A History of the Conquest of Nature in Australia*, North Sydney, Allen & Unwin, 1991, p.223.

13   Richard Flanagan, 'Return the People's Pedder!' Cassandra Pybus and Richard Flanagan eds, *The Rest of the World Is Watching*, Sydney, Pan Macmillan, 1990, p.200.

13   Richard Jones on Lake Pedder: Pybus & Flanagan, p.126, Richard Flanagan, 'Masters of History'.

14   Bob Brown quote: Pybus & Flanagan, p.253, Bob Brown, 'Ecology, Economy, Equality, Eternity'.

14   Richard Jones quote: Pybus & Flanagan, p.209, Richard Flanagan, 'Return the People's Pedder!'

15   Charles A. Reich, *The Greening of America*, New York, Random House, 1970, p.1.

17   Bob Brown on preliminary work on dam: Rosemarie Milsom, *The Age*, 'I Was There', 16 July 2000, p.35.

18   The Wilderness Society had evidence to suggest phone tapping: Milsom, 16 July 2000, p.35.

19–20   Phillip Howard, interview with the author, 13 July 2002.

21   Robin Gray on blockaders: James McQueen, *The Franklin: Not Just a River*, Melbourne, Penguin Books, 1983, p.29. The average age of arrested blockaders was twenty-eight. Students made up 27 per cent of the total, 11 per cent were teachers and 5 per cent belonged to the medical professions. Among almost 1,300 people who had been arrested, only forty-eight were over fifty; see Peter Thompson, *Bob Brown of the Franklin River*, Sydney, Allen & Unwin, 1984, p.174.

22   Doyle, *Green Power*, p.129.

22   Thompson, *Bob Brown*, p.173.

22   Bob Brown warns Liberals: Thompson, *Bob Brown*, p.175.

23   Graham Richardson, *Whatever It Takes*, Sydney, Bantam Books, 1994, p.259.

24   McQueen, *The Franklin*, p.2.

24   Richardson, *Whatever It Takes*, pp.149–150.

25   Christine Milne quote: Pybus & Flanagan, *The Rest of the World*, p.55.

25   Caroline Mee on Tasmania's south-west, interview with the author, 20 July 2002.

28   Richardson on flying over Tasmanian forests: *Whatever It Takes*, p.214.

31   Richardson on woodchipping: *Whatever It Takes*, p.237.

31   Richardson describes document from Naranda: *Whatever It Takes*, p.239.

32   Greens Party poll results: Hutton & Connors, *A History*, p.228.

33   Richardson on Labor's return: *Whatever It Takes*, p.269.

34   Peter Hay on absence of goodwill towards the Greens: interview with the author, 5 July 2002.

34   Christine Milne, correspondence with the author, 24 September 2002.

35   The Greens as a national organisation: the ACT joined in 1993, Victoria and NT in 1995. The Greens WA declined to affiliate and this remains the case today, although they continue to work closely with the national body. In a recent plebiscite on joining the national Greens, WA members needed 75 per cent and received 73.9 per cent.

36   Bob Brown on the nineties: Mike Seccombe, *The Sydney Morning Herald*, 'A Leader from the Wilderness', 16 March 2002.

37   Drew Hutton, interview with the author, 8 September 2002.

40   Hutton, 8 September 2002.

41   Richardson, *Whatever It Takes*, p.257.

43   Geoff Law quoted in Libby Lester, 'Keeping it simple: Environmental conflict and the media in Tasmania', unpublished conference paper, September 2002.

45   Dame Rachael Clelland quote: PM, ABC Radio, 2 July 1999.

48   Greg Baum, *The Age*, 'The Malthouse Effect', Sport Section, 18 May 2002.

50–1   Lucy Turner quote: Darrin Farrant, *The Age*, 'Lawyers Gather in Forests Fight', 11 May 2001.

54   Simon Bevilacqua, *The Sunday Tasmanian*, 21 July 2002.

54–5   John Freeman quote: Danielle Wood, *The Saturday Mercury*, 'Freeman in Race for City's Top Job', 27 July 2002.

56   All figures quoted in this section have been taken from the website of the Australian Electoral Commission: http://www.aec.gov.au/

60   Data for this table was made up from information taken from the Australian Electoral Commission website.

64   Hutton, 8 September 2002.

65   Hay, 5 July 2002.

67   Alex Carey on business-led campaigns: *Taking the Risk out of Democracy*, Sydney, University of NSW Press, 1995, p.3.

67   Carey on New Right agenda: *Taking the Risk*, p.102.

67   For more specific examples of 'economic education', see Amanda Lohrey, 'Components of the New Patriotism', *Island Magazine*, No. 7, pp.12–14.

69   Cameron Andrews, (NSW President of the Democrats and former staffer of ex-Senator Vicki Bourne), *The Sydney Morning Herald*, 'Whither the Democrats', Web Diary, 29 July 2002.

69   Michael Macklin on amalgamation: Tom Allard, *The Sydney Morning Herald*, 'Elder Pushes Merger with Greens', Weekend Edition, 27–8 July 2002. According to Macklin, a Queensland senator for nine years and for a brief time party leader, a merger with the Greens would be a 'natural outcome'. Merger talks in the past had broken down because of egos, not policy. 'Most of those problems were on our side, frankly.'

70   Hugh Morgan on the Greens: Hutton & Connors, *A History*, p.258.

70   Peter Walsh, *The Australian*, 'Goodbye to Life as We Know It When the Greens Get in for the Chop', 6 August 2002.

70   Claims that Greens jeopardise 'government stability': the current drift to the Greens Party among some activists on the traditional Left will continue to fuel such perceptions. On the other hand one can go back to the formation of the United Tasmania Group in 1972 and find the same degree of vitriol and alarmist rhetoric directed at the former Liberal, Labor and Country Party adherents who made that first break onto new political ground.

71   Hugh Mackay, *The Sydney Morning Herald*, 'Comfort and the Green Vision', Weekend Edition, 10–11 August 2002.

73   Cassandra Pybus: Pybus & Flanagan, *The Rest of the World*, pp.66–7.

75   On Bob Brown during Franklin River campaign: James McQueen, *The Franklin*, p. 35

76   On Brown's 'toughness': Hutton & Connors, *A History*, p.158.

76   Brown as an 'organic' leader: Thompson, *Bob Brown*, p.120.

77   Butikofer on Brown: Peter Wilson, *The Weekend Australian*, 'The Greens Add a Touch of Colour', 28 September 2002.

78   For more on the Brundtland Report, see Margaret Blakers ed., *The Global Greens*, Canberra, ACT, The Green Institute, 2001, p.76.

80   Research on larger share of environmental risk for poor: Bovenkerk, *Environment, Community and Culture Conference*, 3–5 July 2002.

81   George Myerson, *Ecology and the End of Postmodernity*, Sydney, Allen & Unwin, 2001, p.7.

81   On global climate change: Myerson, *Ecology*, p.36.

82   The field of 'natural capitalism': Paul Hawken, Amory B. Lovins and L. Hunter Lovins, authors of the best-selling *Natural Capitalism: The Next Industrial Revolution*, are the delayed intellectual offspring of the visionary H.G. Wells who as far back as 1931 argued for economics as a branch of ecology.

Sylvia Lawson

Petrified forest

Away for a few weeks in Europe, I took *Quarterly Essay* 6 with me. A Belgian friend, an assiduous and sympathetic Australia-watcher, picked it up. It's about the future of the ALP, I said. 'Really?' he said with honest surprise. 'Does anybody seriously think it's got one?'

That was in August, as I write now, waiting curiously for the Jenny Macklin policy review and reflecting on the Cunningham by-election result, it's still a good question. 'New believers', writes John Button, 'are the key to the future'; but on his own account of the party in the present, in what or in whom might they believe? With all Button's tough-mindedness, does he or any of his respondents in QE7 realise the size of the chasm between the ALP and those younger Australian worlds from which such new believers might be expected to come? To the three million voters Labor needs, in Barry Jones' analysis, faction-fixing and union-weightings are irrelevant as issues, however their effects may be felt. Labor needs renewal on the most fundamental philosophic and imaginative levels. I can see some parliamentary members who might contribute to it, but they're not among the leadership.

My comments may be taken as complementary to Susan Ryan's. They come from a position among those intellectual workers identified by John Button as a group whose support Labor has lost. He, if not the myopic poll-watchers, believes it actually matters. It's not only that we're communicators, it's also that the issues which matter prominently and articulately to us are also those which – imaginatively, emotionally, subliminally if you like – move many others. From our sideline positions, we can see the wood for the trees, and it looks like a petrified forest.

From the sidelines

I thank Susan Ryan for her definitive exposition of the way in which factionalism, while it corrodes the ALP as a whole, is starkly and persistently a feminist issue. It's that kind of story which explains why so many of us on the left haven't felt able to join branches, however much our principles might drive in that direction: watching the gifted and experienced candidate – often female, and the one more likely to carry conviction with the voters – sacrificed to the dreary, grinding operations of the incurably patriarchal party machine. Sharon Bird's fate in Cunningham ironically reinforces the point; the gender factor couldn't override that constituency's frustration.

Barry Jones acknowledges that 'Many Labor supporters felt alienated from the official line on refugees, [and] defected to cast primary votes for Greens or Democrats … Many traditional Labor voters disliked the "small target" approach and felt that Labor had gone missing on many major issues such as education, health, the environment and immigration since Keating's defeat …'

As one of that many being recognised in those sentences, I'm amazed by Jones' understatements. Is he being super-polite, in a context where there's surely no need, or is he truly ignorant of the anger and frustration not only on the left, but also among genuine liberals, and even genuine conservatives who are democrats enough to want an adequate Opposition? 'Alienated' isn't the word for our response to Labor's support for a heartless and reactionary government's policy on refugees and asylum-seekers; and on the small-target approach, dislike isn't the word either. We were faced by rank political and moral cowardice in the party we'd always voted for, and what we felt (don't kid yourselves) was unequivocal contempt.

No weaker feelings would have impelled our defections; after all, we were used to voting Labor through gritted teeth. In 2001, our acts in the ballot box were as painful as they were unprecedented – and deliberate. Button quotes an old friend on her decision not to vote Green: 'They can't govern.' No, but nor could the ALP. We'd never had more to vote against, we said; when were they going to give us something to vote for?

In a way it was the wrong question. For and against have everything to do with each other; the force of attack on retrogressive policy and practice is equally the scope of new objectives, the extent of ground to be won back. Barry Jones has partly mapped it in his thirty-item inventory of the social and cultural damage wrought by the Coalition in six and a half years.

But the absence of any real light on the hill – the feeble stuff about rolling back the GST, the failure to push through on Knowledge Nation, the frightened avoidance of issues around race relations, multiculturalism, the republic, human rights – all amounted to collusion; the supposed Opposition wasn't convincingly opposing anything. The outcome was a thoroughly dishonourable defeat, and we're still living in its aftermath.

Button rightly satirises the popular post-mortem version that 'the ALP's demise resulted from the dreaded plague *Tampa* released by the wicked apothecary John Howard and his scheming assistants.' That view was always buck-passing; Labor had lost the election well before the *Tampa* left port in Norway. At its acquiescence in the punishment of the victims, some former supporters were grimly unsurprised; after Labor had knuckled under on flagrantly anti-democratic measures like schools funding policy and the private health-insurance rebate, they said, what did we expect?

David Day is right to focus on the education issue. Labor could have reversed its longstanding complicity in the present deeply unjust system of allocation, worked from the many strengths of Knowledge Nation, and affirmed a new commitment to the excellent foundational principle: free, compulsory and secular. This could have been openly and clearly connected to the problems of the bush and the regions, especially for unemployed, under-educated young people and their parents.

But in Labor's pre-election language, those too were forgotten people, and being forgotten, they're not about to remember that Labor's supposed to be there for them. They've breathed in the popular cynicism for politicians; some will vote Green, some One Nation, some for whichever personable candidate happens to show up sounding halfway plausible, party regardless.

In the big picture

Susan Ryan says 'it would be good to hear Simon Crean talking more about reconciliation and the republic.' Indeed, but I'm not holding my breath. Labor has run scared for years now on those larger, so-called big-picture issues which were supposed to be Keating's downfall. At every point along the way the Coalition under Howard has treated indigenous Australia with hostility and suspicion, that 'for all of us' mantra notwithstanding. Post-Wik, the government threw land management as it affects indigenous people back to the states, arguably contravening the *Racial Discrimination Act*, and winding back the modest, cautious gains made under both Wik and Mabo. Thus, despite Beazley's tears in parliament over the *Bringing them home* report, no visible moves were made by Labor, around the election or since, to reclaim indigenous affairs as major national concerns, to deal with history in issues of land, a national apology and a treaty.

Now, after ten years' work, the document prepared by the Council for Reconciliation and its successor, Reconciliation Australia, is on the parliamentary table.

Predictably, the government gives it short shrift. This just might provoke Labor to remember again what Whitlam and Keating never forgot, and what Malcolm Fraser came to recognise: the situation of Aboriginal Australia is an international issue. It has everything to do with the face we present to the world – but that must be a face and not a mask.

Two years ago at least a million of us had a great time walking those symbolic bridges, and nobody then needed telling that the issues in reconciliation sit at the ethical centre of our future. Labor should really spend more time at the movies; it's not for nothing that the strongest feature films of the past year or so have been *Rabbit-Proof Fence*, *Beneath Clouds*, *The Tracker* and *One Night the Moon*.

The fear of freedom

As for the republic, there's a sense that the question has become positively embarrassing. Granted only a little clear analysis, it shouldn't be; that referendum was shackled by a skilful and deeply hostile prime minister, while the ARM and other republicans inexplicably let themselves be conned. Another day will come for symbolic national independence, and Labor should clear its own head on the matter – unless, of course, independence looks just too frightening a prospect.

Among other things, it would imply becoming a critical ally of the United States, instead of a supine little colony. It should also imply a recovery of respect for the United Nations, and for that corpus of international rules which, with quite substantial help from this country in the past, has provided the world with a system of participatory governance. However imperfectly, however often it fails, the recognition and the apparatus have been there.

Now the Coalition, having flouted numerous treaties and conventions on human rights, threatens to support the present US administration's drive to dominate the globe alone, holding a gun to the Security Council's head. Tony Kevin has spelt it out:

> For Howard's Australia and Bush's America, each side compounds the other's foolish aggressiveness and lack of a broader international and humanitarian vision … Australia is moving into the worst of all worlds. Under Howard, we have trashed 50 years of effort to engage seriously with our region and with the UN system of global rules. Yet we have nowhere else to go. The US will not protect us from 'blowback' from our lengthening list of diplomatic blunders … We have become a country that occasionally tries to speak loudly but carries almost no stick at all, and one that has lost its way in

terms of any generous moral vision of its own society or place in
the world. We present ourselves to the world as a selfish and not
very interesting vassal state, dutifully voting with its senior partner
on all major issues.

— Australian Financial Review, 20 September 2002

The thinking Tony Kevin describes is utterly retrogressive, pre-1945. We
should all be frightened by this accurate summation, and by the apparent ease
with which we've turned from a reasonably honourable, minor global player
into something less, and something worse.

The need for an independent foreign policy was, very properly, high on the
Hawke–Wran list of necessary reforms – surely Doc Evatt's ghost was hanging
over them. But at this moment, even taking note of the cautiously oppositional
stance on war with Iraq, and support for the UN in that special regard, it's hard
to imagine Labor shaping up. They could start with that famous speech, quoted
by Button, in which Chifley said: 'We have a great objective – the light on the
hill – which we aim to reach by working for the betterment of mankind *not
only here but anywhere we may give a helping hand*.' (my italics)

There, as a modest internationalist, Chifley gave us the core of what an honourable foreign policy might be. But it's not a separate question; issues in foreign
and domestic policies come together. Hawke and Wran also recognised the necessity for a 'correct and humane policy' toward refugees; another understatement.
The present situation, the maintenance of concentration camps in the Australian
desert, as the work of both major parties, is a glaring national disgrace. While it
continues, we can't credibly parade the globe on missions of human rights.

Through those questions we address the longer term, the reach of history
forward and back. In any progressive politics, the skill is in keeping them in
view while the immediately pressing matters of public policy get their full due
as well. We can do what's needed to give people more hope, more life-chances,
and to bring down household debt, while at the same time keeping alive a
sense of what's involved in national honour. Keating didn't lose because of his
concerns with reconciliation and the republic, and his ambitions for the cente-
nary of federation; he lost because while those matters were trumpeted, it was
widely perceived that he wasn't listening on anything else.

'Ordinary people'

Recovery of principle means, first, renovating the operative notions of the electorate, the capacities of so-called ordinary people, and that's all of us. The rhetorical

sealing-off of the so-called elites – which are always supposed to exclude the highly privileged speakers themselves, Tony Abbott or whoever – is exactly that: rhetorical, ideological, intentionally and viciously divisive. The assumption is that we who debate aloud are somehow outside the world of common folk; the society of which we speak never includes ourselves, it's out there, down there, an undifferentiated mass on a football field, while we're up in the spectator stand.

One great socialist thinker, Raymond Williams, said, 'There are no masses; there are only ways of seeing people as masses.'

In those ways of seeing, voters are reduced to banality. The problem is that Labor falls into the same groove: they, as much as the Coalition, talk so much in terms of 'your average punters', 'ordinary battling families' and so on. John Button gives himself away on this when he writes as though the possession of a university degree and the 'common touch' are necessarily opposed (QE6, p. 14). 'Ordinary people' do in fact consider and imagine their world in larger terms than is generally supposed in political milieux, and the things which are supposed to matter to the much-demonised chattering classes matter – subliminally, imaginatively, mythically if you like – to just about everyone. Attending to opinion polls, however valid and volatile those may be, is not the same as attending to people. If hardship confines minds and energies to immediate survival, that says nothing about the size of the minds in question; it condemns the system which gives their owners less than a fair go.

Watching and hearing

Along with renovation of its social theory, Labor needs a recovery of imagination. The cultural dimension is rather sadly missing from commentary, and by that I don't mean 'cultural policy' in terms of what Bob Hawke used to call the Yarts. Whatever may be feasibly promised in that direction, it would be great to have a sense that our alternative government was actually watching and hearing those Australias which so actively, incessantly, make music, stories, images, performance. Ongoing argument is indeed necessary, but it might benefit from a break in which everybody gave over, shut up, knocked off, stopped counting, closed down the computers and forgot about the polls. Labor's worriers should spend a few months (as suggested above) at the movies, at the theatre, in galleries, music pubs and concert halls – or just sitting round reading books. They'd find, among much else, some lessons in being truly oppositional.

In her excellent new book *Freedom Ride*, on the legendary anti-racist bus trip of early 1965 round rural New South Wales, Ann Curthoys has interesting things to say about the reasons why her own cohort of eighteen- to twenty-year-olds had the guts and confidence to do it. They were too young then to vote, but

> ... they had a confidence distinctive to their generation – a sense of innovation and adventure. Only a few weeks before, the Rolling Stones had visited Sydney, and seven months earlier the Beatles had made their historic visit, rapturously received by youth all over the country. The music somehow expressed the mood of the age group and the time. This was a generation that produced a New Left beyond earlier Cold War certainties ... interested in challenging consciousness as well as social structures and institutions ...
> – *Freedom Ride: A Freedom Rider Remembers*; Allen and Unwin, 2002, p. 37

Then as now the music mattered, both personally and politically. Now the figures of refugee and asylum-seeker are represented over and over in galleries and in eloquent, lively fringe theatre. Perhaps there've been a few party activists in the audiences in Marrickville for the Sidetrack company's extraordinary, wrenching *Citizen X*, the play made from the letters of actual detention-centre inmates.

I'd also like them to see, absorb and think about one of the best Australian films in decades, Rachel Perkins' *One Night the Moon*. This, like *Rabbit-Proof Fence*, tells a story from the early 1930s. A small girl goes missing from an outback homestead; offered the services of a black tracker (who is also a policeman) the father refuses; he won't have Aborigines on his property. Days of searching go on, until the desperate mother goes to the Aboriginal man; they walk into the landscape together, and eventually find the child's body. The extreme simplicity and concentration of cinema strategies – little dialogue, some replaced by singing – makes the ironic, heart-rending story an allegory; the child comes to stand for much else that might have been. At the end we're looking at her grave: the grave of a great opportunity, which didn't have to be lost. We must move on from there, but first it's important to know exactly where we are.

Sylvia Lawson

Chris Ballard

During the 1870s New Guinea was a favoured location for the fictional exploits of romantic adventurers who travelled, much as *Quarterly Essay* Editor Peter Craven writes in his preface of John Martinkus, with danger breathing down their necks at every turn. As these fanciful explorers penetrated deeper into the interior of New Guinea, they encountered ever-larger beasts, more beautiful birds, higher mountains and smaller people – pygmies – inhabiting a fertile and gold-rich paradise. Such was the power of this imaginary that when the first European explorers really did reach the mountains in 1910, in the area where the giant Freeport copper and gold mine is now located, they fully expected to meet pygmies. The people they met in the mountains were indeed short, due probably to the soil deficiencies characteristic of the wet southern fall of the Central Range, and the British explorers duly pronounced them to be pygmies. The descendants of these Amungme people react with a mixture of derisive laughter and anger as they leaf through the photographs of their ancestors taken by these early scientific expeditions; amused by the ignorance of the naturalists, but bitter as they reflect on the lack of change in their condition and on the missed opportunities for benefits or a genuine partnership with successive governments and resource exploiters.

In many ways, New Guinea remains more vivid as a backdrop to the imaginations of distant observers (whether in Jakarta or Australia) than it does as a material location peopled by real communities with very real aspirations. During the short-lived 'Papuan Spring', which extended from the fall of President Suharto in May 1998 until the return of political repression in December 2000, a communicative space, a hole in the ozone layer of secrecy which blankets Papua, opened briefly and Papuan leaders were able to express some of those aspirations directly to the outside world. In February 1999, in the first experiment at direct communication with senior government officials, a team of 100 Papuan civilian leaders met with President Habibie, House Speaker

Akbar Tandjung and the army's Chief of Staff, General Wiranto. The Papuan team respectfully presented their list of requests to the President, the first of which was a demand for immediate independence. Members of the team can still recall the jaw-dropping shock on the faces of the government officials. As a strategic gambit, the Papuan team's opening bid now appears less than wise, for it galvanised Jakarta into concerted action, leading to the diplomatic and security offensive on Papuan separatism which is still unfolding.

John Martinkus's essay cuts through some of the mythology surrounding Papua and, based solidly on a series of sensitive and very useful interviews, gives us a taste of the first fruits of this offensive: the assassination of the civilian leader Theys Eluay, the compilation of detailed plans for the suppression of the civilian movements contained in the 'Matoa' document and the well-coordinated insertion of the Muslim extremists of Laskar Jihad into the western port towns of Papua such as Sorong and Fak Fak. He's at his best when he reports first-hand on his interviews with Presidium leader Thom Beanal, the Dani students at Abepura, US Ambassador Ralph Boyce and the Laskar Jihad members in Sorong. We're given a powerful sense of the depth of Papuan resentment and frustration, and of the gulf that exists between Papuan aspirations and the new fusion of Indonesian nationalism and Islamic fundamentalism. What Martinkus struggles with – and this is true for all writers on Papua – is the challenge of situating these snapshots within an intelligible account of broader processes of change, and then identifying the scope for future movement of the various interests at play in Papua. Once he moves from the visible interaction to the credible report, and then from the credible report to the probable explanation, the picture on the screen starts to break up. In many ways the interviews themselves are more real than the 'facts' or the 'history' which he marshals as context for the interviews.

Numbers are particularly tricky in Papua, and the faithful documentation by Martinkus of the figures recited to him in interview or which he has gleaned from published sources needs to be qualified at every step. Take the question of the population census. The Dutch never succeeded in establishing a total for the population of what was then Netherlands New Guinea, and successive censuses under Indonesian rule have moved increasingly into the realm of speculation (as they have in neighbouring Papua New Guinea). As a rather world-weary local district head or *camat* observed to me when I questioned him about the real figures for the population in his district, 'The village heads lie to me, I lie to the Regent, the Regent lies to the Governor, and the Governor lies to the President.' During a recent census, the enumerators failed even to visit most of the villages in the area where I was working. Estimates of the total population, and of the proportions of indigenous Papuans and non-indigenous residents in Papua, fluctuate wildly.

If it's hard to determine the numbers of the living, it is even harder to be sure about the figures being bandied about for the number of people killed. A macabre 'numbers game' has been played out over the years in activist circles in which numbers of Papuan dead circulate in an inflationary spiral from one report to the next, jostling for media attention with the figures from Rwanda, Sudan or East Timor. Martinkus refers to the figure of 300,000 dead quoted by activist networks, but other sources claim that as many as 500,000 Papuans have been killed. In one report, the claim is made that 8,000 Amungme were killed around the Freeport mine after the 1977–78 uprising (based on the records of local missions, it seems that there were only about 7,000 to 8,000 Amungme in 1977). The 'routine' calculation cited by Martinkus of 100,000 dead is simply the most widely quoted estimate available. Peter Tabuni, an OPM fighter at Mathias Wenda's camp in PNG, gives some suspiciously precise figures for deaths in the most strife-torn of the regencies, Wamena. If we take these figures at face-value (and there would be strong doubts about their accuracy), we get a total of 6,508 from the onset of major conflict in 1977 until the present. Multiply this figure by the number of regencies (though most other regencies have experienced little of the conflict of the Wamena area) and you get a total of 71,588. To this one would have to add the casualties of conflict between 1962 and 1977, perhaps but not probably as many again, yielding an overall total of about 140,000 Papuan dead between 1962 and 2002. Not 10,000, but not 500,000 either.

Why is it important to engage in such a grim debate? In the one area where intensive investigation has taken place over several years, around the Freeport mine, the total number of victims whose identities can be confirmed by the communities of the area and by local OPM units stands at about 220 for the period 1974–2001. This figure certainly understates the real extent of losses, as not all of the dead have been named and those who died of hunger while seeking refuge in the forest are not included, but it gives some sense of the real scale of the tragedy in one of the areas most affected by violence since the 1970s. If the result is less dramatic than the figures doing the rounds on the internet, it is no less tragic for the families and communities from which these 220 individuals were lost. Casually adding zeroes to their number diminishes and effaces the individual tragedies of their deaths just as certainly as their killers routinely seek to remove all trace of the bodies.

At least there's some degree of interest amongst the Western media in Papuan casualties. In Aceh, Indonesia's other province with ambitions of independence, an estimated average of seven people are being killed every day. Where Papua's university rectors can still engage in discussion with government officials about autonomy, and Papua's human rights activists can still conduct parallel investigations into killings, in Muslim Aceh university rectors and human rights activists have long been targets for assassination. This is not a story, however, that plays strongly in the West and the reasons for this have to do with some very murky and unstated assumptions about 'race' and religion. Bylines in international media reports on Papua commonly observe that the Papuans are Melanesian Christians, distinct from other Indonesians on both racial and religious grounds, as if this were sufficient to account for Papuan aspirations for independence. Martinkus and Craven both insist that 'the West Papuans have always had more in common with their neighbours in Papua New Guinea than they do with Indonesians.'

It was Alfred Russel Wallace, the moon to Charles Darwin's sun, who made the first attempt to draw a neat line between 'pure' Malay and Papuan 'races' in the 1860s, but his own writings betrayed the struggle that this required with the evident phenotypical complexity of eastern Indonesia. There are no clear breaks in language or ethnicity as one moves from Papua towards eastern Indonesia. There are indigenous Papuan language communities in Alor, Timor and in the North Moluccas. The western margins of Papua have long been oriented towards the sea rather than the land, participating in Moluccan economies and cultures as much as with other areas of Papua. Some indigenous Papuan communities have been Muslim for almost five centuries – long before most other Papuans became Christian. Indeed Thaha al-Hamid, the outspoken secretary-general of the Papuan Presidium, is a Papuan Muslim. There is also greater linguistic and ethnic diversity within Papua than across the rest of Indonesia, and Papuan leaders are now concerned that their greatest challenge is to achieve reconciliation amongst Papuans. The spectre of an unravelling Papua New Guinea figures prominently in their minds.

This is not to say that 'race' is not an important element in the Papuan conflict. Almost any Papuan can testify to the experience of racist treatment at the hands of other Indonesians. Resentment about such treatment generates a distinctive platform for Papuan resistance. Echoing the earlier racial hierarchies nurtured by the Dutch colonial administration, a common observation made by other Indonesians is that Papuans are 'not ready' for autonomy or are 'insufficiently trained' to staff the administration of their own province. After forty years of Indonesian education and training, the implication is that Papuans will never be ready to administer themselves. This fundamental discrimination against Papuans does add an extra dimension to the conflict in Papua that is not present either in Muslim Aceh or in the Christian areas of the Moluccas, and accounts for the willingness of Papuans to assert a pride in the differences that others insist upon. But to place 'race' at the heart of a new Papuan identity would simply be to replace one form of racism with another.

John Martinkus writes within what is now an established tradition of foreign journalists seeking access by any means possible to Papua and working under conditions that are increasingly dangerous for them and for those Papuans who work with them. A few years ago, foreign journalists and others found to be working illegally in Papua were simply deported, but with two foreign researchers currently in jail and facing trial in Aceh, the prospects for future reportage of this kind are grim. If Indonesian diplomats and security officials are constantly frustrated by what they regard as disinformation emanating from Papua, they have only themselves to blame. Closing the door on Papua will only encourage further speculation on the part of foreign journalists, and possibly a return to reliance on the OPM for information about events within Papua. But whether the Indonesian government can put the lid back on communication from within Papua is now a moot point. A wide range of new and highly articulate sources of information are now available from within Papua, generated by church groups, NGOs, students and researchers.

OPM commander Mathias Wenda's *ennui* at being confronted with yet another journalist making the pilgrimage to his camp is evident, and Martinkus is professional enough to convey this. Not surprisingly, the results of this interview are the least interesting, and the least well-informed of those in this essay. Most of the senior OPM commanders have been in the jungle for a quarter of a century, since the 1977 uprising, and their sense of Papuan, Indonesian or global events is inevitably restricted as a consequence. They have played an important role for many Papuans in the past, and they are certainly a part of Papua's future, but the OPM are only a tiny fraction of the myriad of different groups in Papua, each of which has different interests and different approaches, albeit in pursuit of a broad common goal. Particularly disappointing in Martinkus's essay is the near total absence of voices from any of the churches (one local pastor is brought on to talk about the Laskar Jihad). Christian church and Papuan Islamic leaders play a far greater leadership role in Papuan communities than either the OPM or the Indonesian government. There is increasing discomfort within Papua at the focus placed on OPM commanders by foreign journalists, and at the romantic image of the OPM as spokesmen for all Papuans, promoted by editors musing on the OPM's 'quixotic nobility in the face of futility' (to quote Peter Craven in his breathless preface to Martinkus). Papuans need and deserve rather more than this. What the army currently fears far more than the OPM – who have never posed a problem militarily – is the church-sponsored and Presidium-sanctioned 'Zone of Peace', which calls on all sides to refrain from conflict, a move which threatens to expose those parties with an interest in sustaining the violence.

Perhaps the most significant casualty of this international media focus on OPM opinion is the perception – which is widely held within Papua – that there is no space for engagement between the polar opposites of Indonesian nationalism and Papuan independence. Martinkus is accurate in his portrayal of widespread Papuan scepticism and suspicion in response to Jakarta's offer of special autonomy, but his dismissal of the Papuan government officials, academics and church leaders who have been trying to promote discussion about autonomy as 'precisely the groups that stand to gain financially' is both naive and grossly unfair. Advocates of engagement in the autonomy process acknowledge that Jakarta's entire approach to the autonomy legislation has been dismissive of Papuan aspirations. Who would have expected otherwise? Yet the fact remains that even the limited concessions of the autonomy bill potentially mark a major departure from thirty years of a largely repressive relationship. The autonomy process – and it will be a process, not an overnight phenomenon – represents the only space for discussion and negotiation between Papuans and Jakarta that is currently on the horizon. What Martinkus fails to report is the very deep antipathy and suspicion of autonomy being voiced openly by Indonesia's nationalist politicians and security chiefs who regard it as the first step towards independence, and who have no intention of making even the limited concessions outlined in the bill. The autonomy bill got through parliament in Jakarta only because of the persistence and perseverance of the Papuan drafting team, who spent months in Jakarta meeting politicians and promoting their cause, and because parliament was in the process of impeaching President Wahid when the bill was brought before them.

Where should outsiders – and, in this context, Australians – stand on the Papuan conflict, and how should we express our views to best effect? First up, Papua is not 'the next East Timor', as every second foreign pundit proclaims. This is tired and lazy 'soundbite' analysis (why, incidentally, is Aceh never the next East Timor?). Quite apart from the differences between Papua and East Timor in terms of history and legal status, what Papuans need is sustained engagement with the intricate details of their past and present situations rather than this airbrushing of political complexity. Martinkus actually works quite hard to register the subtlety of a position such as Thom Beanal's – a leader caught between multiple constituencies, who is at once a tribal chief, a commissioner on Freeport's board and the acting chair of the Papuan Presidium. But Craven's preface slips into stereotype in his mild condemnation of Beanal as 'far more temporising' than the heroic Theys Eluay, 'the one figure around whom the independence movement clustered'. Anyone aware of the actual events at the Second Papuan Congress which saw Theys elected as Chair would find this hilarious. Theys Eluay certainly emerged as a symbol of the Papuan Congress movement, and his assassination has united Papuans in their demand for his true killers to be brought to justice, but a biography of Eluay in the June edition of *Inside Indonesia* indicates some of the ambivalence with which he was regarded by many Papuans for much of his life.

To my mind, it's both inappropriate and irresponsible for foreigners to call for independence for Papua. Inappropriate (and largely counter-productive) because we aren't Indonesian citizens, and irresponsible because we won't have to live with the consequences. It *is* appropriate and it *is* a responsibility for us to work for basic human rights – for adequate political representation, the right to self-determination, and freedom from violence – for Papuans, for Acehnese, for all Indonesians, along with Aboriginal and Torres Strait Islander people and refugees in Australia, among others. Papua's most immediate problems are essentially Indonesia's problems, chief among which are its mercenary army and ineffective judiciary. If Martinkus's essay establishes one thing, it is that the rest of Indonesia, and the wider world, knows very little about Papua. We need more journalists on the ground in Papua, and more interviews with a wider range of perspectives, Papuan and Indonesian. And we need, in particular, a journalism which opens up new forums and spaces for dialogue, rather than closing them down by focusing on the voices of polar opposition.

Chris Ballard

Bruce Grant

John Martinkus raises issues of consuming human interest and increasing diplomatic importance. His essay is rich in detail and useful references. His account of the American ambassador's visit to Sentani in April this year, with its mix of formality at the grave of the Papuan leader Theys Eluay, murdered last year after a dinner meeting with officers of *Kopassus* (Indonesian Special Forces), raucous student enthusiasm, bloodcurdling war whoops, suppressed violence and the presence in everyone's minds of the huge Freeport-McMoRan gold and copper mine, Indonesia's largest single taxpayer, is a brilliantly suggestive cameo of the real world in which the drama of West Papua is being played out.

What remained in my mind when I had finished reading was not just the maltreatment of the Papuans he documents, but their predictable behaviour in response, in particular their devotion to a flag and the use of physical violence in their just cause. No one can blame them, any more than one could blame the East Timorese (or the Palestinians and many others) for wanting a state of their own and a flag of their own to fly as a symbol of their independence, and being prepared to fight for it. It is the way of the world, expressed in the adage tattered but still resonant, that 'states make war and war makes states.' But the association of military power with the sovereignty of the state is the very combination that makes the abuse of human rights, including that in West Papua, so prevalent.

The system of states that has operated in the modern world since the Westphalia treaties of 1648 is based on the control by those states of borders and a stable population. Without control of their borders and their citizens, states are subject to interference from outside by stronger states or by non-state forces. When the United Nations was formed in 1945 with the states as its members, non-interference in the internal affairs of states was one of its principles. But another of its principles was that every person had rights, which were announced in the Declaration of Human Rights three years later.

It was evident even then that the right of the sovereign state to control its citizens and the human rights of those citizens were likely to conflict. At the time that Indonesia took over West Papua (then Irian Jaya) from the Dutch, the rights of the state were superior, in world politics and in international law, to the rights of the individual. Since particularly the end of the Cold War and a revival of ethnic and religious identity, accompanied by the surge of globalisation, the clash between the rights of states and the rights of people has become epidemic. The state and its instrumentalities, the police and the armed forces, are the greatest abusers of human rights.

In 1999, in a far-reaching speech to the UN General Assembly, Kofi Annan assessed the implications, describing it as a new era of 'two sovereignties', the sovereignty of the state and the sovereignty of the individual. First, he said, state sovereignty was being redefined. 'States are now widely understood to be instruments at the service of their people, and not vice versa.' Although states were the foundation members of the UN, the fundamental freedom of each person was at the core of the UN charter and subsequent international treaties. 'When we read the charter today, we are more than ever conscious that its aim is to protect individual human beings, not to protect those who abuse them.' The world – and therefore the UN – could not stand aside when gross and systematic violations of human rights were taking place. Intervention, however, must be based on legitimate and universal principles. In a thoughtful, penetrating address, he tackled the need to redefine not just 'sovereignty', but 'national interest', 'vital interest' and 'common interest', terms that have been used glibly, as if we all understood what they meant.

At the same time, international criminal law has been developing, lately rapidly, as ad hoc tribunals are set up to deal with massive human rights violations in situations of conflict and, most recently, with the advent of the International Criminal Court (ICC). The trial of Slobodan Milosevic for crimes against humanity allegedly committed while he was Serbia's head of state is merely the most noticeable of a number of prosecutions of agents of the state. The ICC, established this year, will for the first time provide a permanent international forum to prosecute individuals as well as states. If a soldier behaves badly in contact with the civilian population, he or she could find himself before it. For the first time recently, Serbian soldiers were convicted and gaoled for rape. This is the new reality of world politics and international law in which the West Papua issue needs to be resolved.

John Martinkus has written a gripping narrative, without an ending. He assumes at times that West Papua will, or should, end up in one way or another like East Timor. However, West Papuans in the separatist movement (OPM) face a much more difficult situation, not the least of which is the example of East Timor itself. Also, since 11 September 2001, another tattered adage – 'one man's terrorist is another man's freedom fighter' – has been given a new lease of life, with states as politically different as Russia, China, India and Israel united in their determination to stamp out their particular brand of terrorism. What emerges from Martinkus's account is that the Indonesian military is playing a more subtle game in Papua than in East Timor, framing the Papuans as the terrorists and, in one of many ironies, importing into their midst elements of Laskar Jihad, the Muslim militant organisation, to stir up trouble. They are aware, like everyone else, that the public's mood at the moment, and the disposition of governments, is to have no sympathy with terrorism, whatever its underlying causes may be. The Bali massacre will overshadow Indonesia's internal security problems for a while, but the West Papuan issue predates current terrorism concerns and will continue however they are met.

The cases of West Papua and East Timor although similar in their experience of human abuse, are different in other important respects. East Timor was Portuguese, and therefore did not pass to Indonesia at independence as part of the Dutch East Indies. It is true that the Dutch withheld Papua from Indonesia, the successor state, until the 1960s, but it was part of the Dutch East Indies and its passing to Indonesia was as natural and appropriate as the artificial terms of 'state' and 'nation' allow. It is true that the UN-supervised Act of Free Choice was an unhealthy mix of pyramid selling and branch stacking, but Indonesia was not a democracy at the time and what the UN arranged was, in the circumstances, better than the alternative on offer, an Indonesian military assault. In addition to these differences, the humiliating fact of East Timor's independence hangs over any future Indonesian government, especially its military component. When the current president's father was president, he used to proclaim the unity of Indonesia from 'Sabang to Merauke', or from its western to its eastern tips. This did not include East Timor, but it certainly did include Irian Jaya, and the sensitivities of Indonesian nationalism would be aroused much more by the threatened loss of West Papua than they were by the loss of East Timor. The sensitivity of Indonesia's budget bureaucrats would also be aroused: West Papua is rich in resources.

In Australia, the cause of East Timor's independence was sustained by a campaign to avenge the deaths of Australian journalists at Balibo during Indonesia's military intervention in 1975. The Australian media kept a particularly watchful eye on the situation in East Timor. In addition, East Timor's resistance leaders had a European, even romantic, style that Australians like and which the Papuan warriors, with their authentic aboriginality, cannot replicate. Even so, many Australians (myself among them) supported Australian policy until the massacre in Santa Cruz cemetery in November 1991. After that, realising that the Indonesian military was incorrigible, many still preferred a form of autonomy for a period of perhaps five years before a vote on self-determination could be taken. As it turned out, of course, the people of East Timor were given their opportunity to be independent in quite different circumstances. The catalyst, however, was not the Santa Cruz massacre or many other violations of human rights but the Asian financial crisis of 1997, which destroyed the Suharto regime and its firm, military hold on the Indonesian archipelago and brought to office Suharto's vice-president, who was not at his best when confronted with complex political issues, including a metastasising military.

The treatment of the Papuans, like the East Timorese before them – and like the Acehenese – could become a source of instability in the region. But Indonesia is fortunate. Unlike Yugoslavia, it is surrounded by countries that do not want to dismember it. When Slovenia and Croatia put up their hands, Germany and then most European and Western states, including Australia, were quick to recognise their independence. When the Cold War ended, Yugoslavia was a federation of six republics. Only two are left.

Indonesia is as unnatural a nation as Yugoslavia was, hastily put together as it was after the First World War. Like Yugoslavia it is on a crossroads, in its case a cross-waterways. In Asia, only India is more culturally diverse, but India has a natural unity created by mountains to the north and sea to the south. Indonesia is an archipelago of thousands of islands straddling a maritime freeway between the Pacific and Indian Oceans. Its outer islands are rich in resources and resent the dominance of over-populated and introspective Java, just as Serbia's dominance was resented in Yugoslavia. Aceh has always been recalcitrant, Sumatra (led by some dissident colonels assisted by the CIA) tried to break away in the 1950s, the Moluccas (the Spice Islands) have a resistant Christian community in the south, as did East Timor and as West Papua has now.

Indonesia and Yugoslavia were both non-aligned during the Cold War. When, after Tito died, the Cold War declined and Yugoslavia's neutrality became irrelevant, Milosevic tried to enforce Serbia's dominance with catastrophic results, including his own removal to The Hague. Indonesia was held together during the Cold War by an understandable nationalism after centuries of Dutch colonial rule, expressed by Sukarno, and, in Suharto's time, by economic development and military discipline. It is possible that the third president, B.J. Habibie, a technologist, would have been a moderniser, connecting the country with globalisation, but he did not stay long enough for us to find out. The fourth president, Abdurrahman Wahid, had a sense of the complexity of the contemporary world and the need for Indonesia to become more flexible and cosmopolitan in order to find an accommodation with it, but he did not last long enough either. Now we have Sukarno's daughter Megawati who seems by temperament inclined to the nationalism of her father and to a degree to the support of the military, but the experience of Yugoslavia — and Milosevic — must give her pause.

She cannot rely on her father's message of nationalism as a balm to soothe the poverty and suffering of the people, nor on Suharto's military repression for the benefit of a plutocracy. The prospect is that the OPM will continue to accumulate arms and turn to violence as a way of keeping its claims alive, the Indonesian military will continue to use counter terrorism to justify its presence and journalists like John Martinkus will continue to bring the depressing results to the outside world. The Indonesian government — and people — will continue to suffer from an unsavoury reputation that has dogged the republic, especially since East Timor, but before that, going back to the massive slaughter of 1965–66 and including the mysterious (*petrus*) killings in 1983, for which Suharto in his autobiography took responsibility, saying they were criminals, the Tandjung Priok riots in 1984, the murder of the young factory worker Marsinah in 1993, just to mention a few that happen to come quickly to mind.

Wahid's idea of a new regional group, South-West Pacific Dialogue, is a useful beginning. The membership — Indonesia, Australia, Philippines, Papua New Guinea, New Zealand and East Timor — was designed to offer East Timor access to the region without throwing it immediately into the larger ASEAN pond, but at its first meeting in October 2002 the main topic seems to have been West Papua. The original meeting was to be held in Papua but was moved to Jogjakarta because of the murder of schoolteachers from the Freeport mine. The rhetoric at its first meeting in Jogjakarta in October 2002 was predictable, Indonesia declaring that anything more than the limited economic autonomy proposed for West Papua was impossible and Australia declaring that it had no wish after East Timor to see Indonesia further dismembered. Their positions are understandable. Apart from the politics of Indonesian national pride, Australia has enough on its plate, both financially and militarily, with East Timor, to make uninviting the political consequences of another conflict with Indonesia. Neither East Timor nor Papua New Guinea showed any support for the OPM. A New Zealand offer of mediation was rejected by Indonesia.

Challenges, however, create opportunities. It would be a welcome break in this depressing story for President Megawati Sukarnoputri to turn the problem of West Papua into an opportunity to show that Indonesia is moving into the twenty-first century with an identity that is more sophisticated than that which, in the name of nationalism, claimed sovereignty over West Papua. If Indonesia would really like to 'startle the world', as Sukarno often claimed he was about to do, it can honour its promise of autonomy for West Papua by giving it political as well as economic substance. The first step would be to break the sterile tension between the Indonesian military and the OPM. One way of doing this would be to accept the need for an international presence as a buffer between the local people and the Indonesian police and armed forces. A token international police presence should be all that is necessary and the United Nations has the experience to undertake such a sensitive operation. Its civilian police activities, although small, have grown each year in the past decade (from 1,053 personnel in 1993 to 6,754, in seventeen missions, already this year). It is interesting that the United States, which is not a major contributor to UN military peace-keeping, is, with India and Jordan, active in civilian policing. Indonesia's need for American financial and political support makes the American position important, especially as the Freeport mine, like the Bougainville copper mine in Papua New Guinea before it, seems likely to be a target of the segregationists.

The point of international intervention would not be to foreshadow independence for West Papua but to give autonomy a fair chance to prove itself as a viable option. From Indonesia's point of view it is an opportunity to establish internal order in the province *before* rather than *after* the next unsavoury incident occurs. International policing of internal order is a solution that is rarely considered until it is too late, when the state has lost control, law and order does not exist, human rights violations become massive and the more dramatic remedy of international military intervention has to be considered.

Pre-emptive policing takes the heroics out of conflict, which is why it is not favoured by those who are committed to the use of military force as a solution to the conflict of rights and interests. In this case, the disenchanted are the Indonesian military and the OPM. But, from the point of view of the Indonesian government – and, indeed, the Indonesian people – it is worth considering. Indonesia's neighbours need to impress on it that they are willing to co-operate to find a solution for West Papua, but that the Indonesian state needs to loosen its military hold on the archipelago for this co-operation to be effective.

Bruce Grant

James Griffin

What is he whose grief
Bears such an emphasis? ...
... Dost thou come here ...
To outface me with leaping in her grave?

Hamlet, Act V, Scene ii

'*Paradise*'? Such a soggy cliché. I'm surprised at Peter Craven. 'Paradise' should have gone out with Dorothy Lamour, or at least post-war with Michener, Rodgers and Hammerstein. To recall a soggy song[1], pre-colonial evenings in 'West Papua' were more likely to be ensorcelled rather than enchanted. 'Paradise' can also make for sardonic comment not just on the long term rapine of imperialism but on the frailties of post- and neo-colonial so-called 'inheritance elites'.

And '*Betrayed*'? Perfidy? A sell-out of some firm obligation? I seem to remember radicals of the fifties rejoicing that Melanesians in West New Guinea might escape the sort of white man's bondage imposed in the East, in the Territory of Papua and New Guinea, TPNG as it was known.[2] Arguments from ethnology, cartographical rationality and even bio-diversity that West New Guineans were a distinctive people with a non-Indonesian destiny did not persuade those who told us how they would be accommodated by the shibboleths of *Unity in Diversity* and *Panca Sila*. Were the Dyaks of central Borneo, they asked, complaining of their slower ride to modernity? What was good enough for them was surely good enough for West New Guineans.

However, Craven suggests there ought to have been something almost proprietorial in the relationship between Australia and New Guinea in its whole cartographical sense. 'This is our back door, what should be our sphere of influence, on which the severed body parts fall,' he says, without a blush of chauvinism at this recap of our nineteenth-century pseudo-Monroe Doctrine. In reality West New Guinea looks more like Indonesia's back door than a site for an

Australian jerry-built outhouse. In historical terms it clearly had a plausible claim. A little history can also help with such gung-ho analogies as

> ... West Papua is another East Timor waiting to happen ... in fact it is happening with the collusion of Australia [*sic*] and American indifference.

Here I take it that 'Australia' is not an epithet and that it has been its government and the abstract noun that have somehow 'colluded' and that therefore it may have been even worse than indifferent. In haste to apportion guilt Craven has muddled his syntax.

While I would not question the veracity of John Martinkus, it is a pity he is not more precise about his itinerary and the amount of actual time he spent in what, optimistically, he chooses to call 'West Papua' rather than its present official name, Papua. I also wonder if he spoke only in English in his interviews and how well his interlocutors replied. This is not carping when a journalist claims to know that 'the general community is cynical' about Jakarta's offer of autonomy to the province. Not that anything that Martinkus writes lacks credibility; in fact, even his self-conscious intrepidity is rather déjà vu. His credulity, however, is untested. Unfortunately there is no sense of any of this in Craven's 'beat-up' introduction with its unwarranted hyperboles (e.g. 'absolutely level and absolutely convincing') and that fatuous adjective 'riveting' which do no service to the subject nor credit to a distinguished literary critic. Neither Martinkus nor his editor prepared himself thoroughly for his task. A journalist who believes that Australia 'made the border' between Papua New Guinea and 'Papua' or, to be more instructive, between Papua New Guinea and Indonesia or, to be portentous, between the Southwest Pacific and Asia, is hardly fluent in the history of the great 'bird-reptile island'. Nor is one who thinks Sukarno first 'laid claim to West Papua' in 1954 instead of at least 1945, before *Merdeka*. And any one who locates Kiunga as a 'village south of Vanimo' is hardly precise about its geography.

The Commonwealth of Australia did not exist when the tri-partition of New Guinea between the Netherlands, Great Britain and Germany took place. The Dutch had claimed territory on the north coast up to the 141st meridian as early as 1828.[3] Extending this line southwards (notably excluding the westward bulge of the Fly River) conveniently divided the island into roughly equal parts and was accepted by Great Britain and Germany which, drawing a couple of straight lines through the unknown eastern Highlands, acquired a quarter each. Australia in 1906 took over in sovereignty British New Guinea as the Territory of Papua and in 1921, following conquest in 1914 by our 'Coconut Lancers', acquired a C-class mandate over (formerly German) New Guinea from the League of Nations. Pre-war the Dutch took little interest in their New Guinea possession, setting up remote administrative posts at Manokwari and Fak Fak only in 1898 and at Merauke in 1902. Perhaps 200,000 were under control in 1937, less than 30 per cent of the population. There was some contact between West New Guineans and other inhabitants of the Netherlands East Indies (NEI): outside the major centres minor official and missionary posts were filled by the latter while Indonesian political prisoners, who from 1927–8 were marooned 500 kilometres up river from Merauke, acquired a sense that West New Guinea was an intrinsic part of the state they hoped to inherit. There were intimations that it could be rich in oil and other minerals but not such as to generate dreams of El Dorado.[4]

Prior to the end of World War Two serious discussions took place among Indonesian leaders as to whether the borders of their new state should be NEI minus West New Guinea (a position taken by future Vice-President Hatta) or NEI including West New Guinea, or *Indonesia Raya* including Malaya, all territories of Borneo and even East New Guinea and the Solomons (based on fantasies of the medieval Srivijaya and Madjapahit empires). A clear majority opted for the borders of the colonial state, a principle generally observed elsewhere at decolonisation. In 1949–50 the Netherlands appeared to agree to this but for domestic political reasons and, in the hope of protecting investments in Indonesia, changed course.

A review of Australian policy of the time is depressing because we imagined we were free to act in our 'strategic perimeter' without subordination to the UK or the USA. Assuming that we had become some regional 'middle power', the ANZAC pact of 1944 had asserted a right to be consulted in any final settlement there. When External Affairs Minister Percy Spender in August 1950 rejected Indonesian claims in a major statement at The Hague, he stressed the strategic importance of West New Guinea to Australia, and expressed fears that other claims would not end there, and that Communism was spreading. (China had become Communist in 1949, the Korean War had just begun and falling dominoes were the current fantasy.) Little consideration was given to the political rights of 'West Papuans'. Labor's Dr Evatt at one stage suggested that Australia might purchase West New Guinea from the Dutch. (Ironically, years later in Papua New Guinea vociferous anti-colonial students expostulated that that was what should have been done!) When R.G. Casey succeeded Spender in 1951, he put the issue into 'cold storage', as he said. He hoped to lull popular feeling and direct attention to what he thought were more urgent problems of stability and development.[5] All very gentlemanly, but unrealistic. President Sukarno was already using the issue as a rallying point for national solidarity and would go on to make West New Guinea a focus for revolutionary goals and a scapegoat for even self-inflicted economic ills.

In 1954 when Indonesia submitted the issue to the United Nations, the Dutch were still relying on legal arguments for the basis of their sovereignty instead of on the right to be prepared for self-determination. That a 'primitive' people should be rushed into independence within ten to fifteen years was still considered absurd. Meanwhile for Australia our security was paramount; self-determination was mentioned but only as a secondary consideration. By the 1955 UN session Indonesia's sense of moral right had been fortified by the anti-colonialist manifesto of the epoch-making Bandung conference where all twenty-nine Third World nations supported Indonesia's claim. New Afro-Asian nations admitted to the UN, like the Communist bloc, were unimpressed by Australia's security fears. When in 1956 Casey emphasised that the 'fundamental issue' was the welfare of West New Guineans, and the Dutch now belatedly called for self-determination, other nations had good reason to doubt their motives. TPNG was showing little sign of devolution. In 1951 when Paul Hasluck took the portfolio of Territories he said it might be a hundred years before TPNG was self-governing. Even the idea of future seventh statehood was being canvassed. A suspicion was unavoidable that Australia had the same designs on its UN mandate as apartheid South Africa had on Southwest Africa (Namibia). In 1959 the future Governor-General, John Kerr, an enlightened analyst for his time, advocated a Melanesian Federation, joining both halves of the New Guinea island, its eastern islands and even the British Solomon Islands Protectorate.[6] It was a tidy and humane scheme but – much too late. If there had ever been a time for some such solution under a UN Trusteeship involving Indonesia, Australia, NEI and the UK, it would have been 1949–50 before Indonesian nationalism gathered momentum.

Two statements in November 1958 expose the unreality of Australia's position. J.F. Dulles, the formidable US Secretary of State, reassured us that, if Indonesia attacked West New Guinea, it 'would have the US to deal with'. The distinguished Australian commentator W. Macmahon Ball could say that our stand on West New Guinea was 'not obscured or compromised by our relations with our powerful friends'.[7] But within three years Indonesia had acquired it with full American, Dutch and British concurrence. There is no need to recall here President

Kennedy's justifiable fear that Sukarno would recklessly move closer to the Communist bloc (which would not have benefited 'West Papuans') unless he triumphed in West New Guinea; or the ostensibly high-minded Bunker Plan which brought it about by mid-1963. In protest, the Australian Leader of the Opposition, Arthur Calwell, foolishly rattled a blunt sabre. Ironically, 1961 was the year of our lowest defence expenditure of the decade and at a level which Labor had still thought too high. Australia now knew it was not even a middle power. It had been shadow boxing well above its weight. To ratify Indonesia's takeover only the Act of Free Choice of 1969 (the 'Act of No Choice') was needed. Under the *musjarawah* consultation 1,025 carefully selected representatives voted on behalf of 800,000 people. It was dubbed *All Men One Vote*. The UN's Bolivian supervisor, Ortiz-Sans, became so elusive as to be called 'Slippery Sans'; he put in a mildly contrite report that a legitimate plebiscite had taken place. Australia's then minister, Gordon Freeth, like most other representatives, approved while maintaining he only acquiesced.

Martinkus, as a result of experience with drunken soldiers at Vanimo, implies that Papua New Guineans are uninterested in the plight of their Melanesian brothers. This is untrue although officials are obliged to recognise Indonesia's sovereignty. One excellent example of ambivalence always comes to my mind even though it happened twenty years ago. When he was in office successively as Secretary for Defence and then Minister for Foreign Affairs (1980–2), the current retiring Secretary of the South Pacific Commission, Noel Levi, upheld Indonesia's sovereignty over then Irian Jaya, as he was required to do, and explored the possibility of full membership of ASEAN for his country. Out of office Levi wrote that PNG 'was not complete without the western half of the island and that accepting the ASEAN charter would endorse Indonesia's claims over West New Guinea, the worst kind of legacy we will leave to future generations'.[8] In mid-1998 the Governor (i.e. Provincial MP) of Sandaun (West Sepik border province) pleaded for the recognition of West Papuan independence. In December 1998 a World Council of Churches office in Port Moresby was shut down by police after an Indonesian embassy protest but the nation's daily newspaper editorialised soon after, 'Listen to the cries of our Melanesian Brothers'.[9] However, Port Moresby is properly in awe of Indonesia's size comparative to its own and has been wary of popular sentiment in this matter. In 1984 some 11,000 refugees (over 1 per cent of the indigenous population) crossed the border into the Fly and Sepik regions overtaxing local resources. Many are still there. There is always a menace of more to come. The Papuan rebels (OPM) can expect only border havens in Papua New Guinea. But pan-Melanesian sentiment is widespread and unlikely to wither.

Many more people are aware of the barbarities committed in Papua than Martinkus and Craven seem to realise. But what can be done? Suggesting that East Timor is a precedent will not be helpful. The difference in pre-independence status may seem minor but will remain crucial for support in the UN. The Act of No Choice will not be reversed. In the fanciful event that a genuine referendum will ever be sanctioned in Papua, transmigrants who today number some 40 per cent of the population will have to have a vote. The logistics will be almost insuperable and any result probably contested in a civil war. Relinquishing Papua after fifty years of nationalistic 'struggle' would flatten Jakarta's morale and lead to separatism in the provinces with unpredictable results for the region. There is also the resource value of Papua to Indonesia. Conceivably, with a supply of arms from somewhere (but where?) the OPM could immobilise the Freeport mine and other projects, but Papua is not Bougainville. Short of some breakdown of the state, Freeport will hardly ally itself with destructive rebels.

East Timor's foreign minister, J.R. Horta, who has witnessed the cost to his people of achieving independence, even with UN support, has advised Papuans to accept Jakarta's current offer of autonomy to which is attached the allocation of 70–80 per cent of provincial resource revenue. How sincere Jakarta will be when it comes to implementation is for conjecture – one can only be pessimistic – but in principle it must be seen as a step in an ameliorating direction, even one that could lead non-violently to virtual internal self-government. However, to Craven and Martinkus autonomy is merely 'a sop'. So what are they recommending aside from guilt? For the historical record, perhaps they can tell us how Australia should have acted and in what way we could have 'liberated' Papua but in what decisive way Australia 'sold out'. The Martinkus essay is an appreciable if limited piece of elevating reportage, but will not, as Craven believes, 'alter the picture of West Papua'. Those of us who have always impotently thought that Papua was potentially a rational political entity, amalgamated with Papua New Guinea or not, and has been denied natural justice, may be none the wiser.[10]

James Griffin

1 A younger generation may need to know that the popular song was 'One Enchanted Evening' from the famous post-war musical, *South Pacific* (1949).

2 An extreme case as late as the 1980s was Papua New Guinea's naturalised novelist Dr John Kolia (Collier) who, after visiting Irian Jaya, wrote: 'I can compare Irian Jaya with PNG and am able to say that it is a happy, peaceful place with the economy in the

hands of its citizens, not of foreigners'. J. Kolia, *Irian Jaya, 1980s: A Writer's Diary*, Aklat Publications, Hong Kong, 1980.

3 P.W. van der Veur, *Search for New Guinea's Boundaries*, Australian National University Press, Canberra, 1966, passim.

4 R. Garnaut and C. Manning, *Irian Jaya*, Australian National University Press, Canberra, 1974, 9–12.

5 G. Greenwood and N. Harper, *Australia in World Affairs 1950–1955*, F.W. Cheshire, Melbourne, 1957, 204–5, and *Australia in World Affairs 1956–1960*, F.W. Cheshire, Melbourne, 1963, 283–9.

6 'The Political Future' in J. Wilkes ed., *New Guinea and Australia*, Angus & Robertson, Sydney, 1958, 138–75.

7 Quotations in J. Griffin, 'Suppression of West Papuans', *Catholic Worker*, June 1969, 6–8.

8 N. Levi, *Times of PNG*, 8 February 1984.

9 *Post-Courier*, 25 June, 14 December 1998, 5 January 1999.

10 The most recent and highly recommended book on this subject is C.L.M. Penders, *The West New Guinea Debacle: Dutch Decolonisation and Indonesia, 1945–62*, Crawford House, Adelaide, 2002, also recommended is S.R. Doran, 'Western Friends and Eastern Neighbours: West New Guinea and Australian Self-Perception in Relation to the United States, Britain and Southeast Asia, 1950–1962', PhD thesis, Australian National University, July 1999.

Andrew Hamilton

The challenge posed by *Paradise Betrayed* has nothing to do with the strangeness of West Papua. It arises from the familiarity of an account of yet another counter-insurgency program. The challenge is to overcome the moral paralysis so easily induced by this depressing familiarity. For experience suggests that whenever the armed forces of a nation are deployed with overwhelming superiority against a distinct group of its own nationals who wish to retain their own culture and links to their land, resistance leads to massacre and repression. Experience also suggests that the oppressed will receive little help from other nations, unless it is in the narrowly defined strategic interests of those nations to offer it. And any assistance that is offered will rarely be for the people's comfort. So, while moral decency leads us into solidarity with the people in their aspirations for a human life and in their resistance to brutality, political necessities forbid us from offering them encouragement.

Paralysed between two paths that lead to undesirable destinations, two unpalatable paths, the easiest way to move on is to throw away the moral compass. Most political commentators, indeed, prefer to analyse the situation in large terms that disregard the human reality of counter-insurgency. They submerge it beneath large theories about strategic relationships and trade. The issues which they find determinative include the territorial stability of Indonesia, the need for a strategic counterbalance to China, the demands of mounting an effective global war against terrorism, the incapacity of a small and primitive nation to exploit its resources, and the inevitable failure of any movement for independence. When the mind is freed to contemplate the abstract beauty of such exalted goals, it can then easily approve the military strategies that are necessary to achieve them, and make little of the murder, rape, assault, extortion and despoliation in which these strategies find expression. These are sanitised and packaged as human rights issues, to be stored at the back of the shop in case someone should ask about them. This approach became familiar in writing over twenty-five years about East Timor, in which we were urged to forget about unfortunate incidents and concentrate on the big picture.

But the East Timorese changed the picture. In doing so, they showed that a political rhetoric that throws away the moral compass simply neglects the human costs of the policies which it is used to advocate. It does not alleviate them. The East Timorese also showed that settlements imposed by terror are not as inevitable and irreversible as they had been told they were. Nor were the consequences of the new settlement as disastrous as had been claimed. The East Timorese showed finally that when we fail to give full weight to questions of justice and human suffering in our political thinking, our rationality will inevitably become a weapon used to support continuing terror.

When reflecting on the counter-insurgency strategy in West Papua and elsewhere, we need to maintain a moral perspective. Although large questions of policy cannot be answered by moral considerations alone, neither can they be answered decently or wisely by considerations of self-interest alone or by abstract theory in which human suffering is disregarded. To find a moral standpoint from which we can ask complex strategic issues is difficult, but John Martinkus's approach in *Paradise Betrayed* embodies elements that should be part of any response.

Before asking questions, he accompanies the people of West Papua, often putting himself at some risk. His example suggests that if we wish to discuss issues that affect the lives of people affected by policy, we should first come to know their lives and become familiar with their faces. When you accompany people, it is also natural to consult them about their experience and their desires. It would seem odd to devise and impose upon them a policy that affected their lives without first consulting them. In democratic theory, the importance of consulting the people is normally emphasised. But it is less honoured in practice, especially in the case of minorities within nations. In West Papua, both the United States and Australia supported the rigged plebiscite of 1969. The people were not consulted. In a society founded on such a fraudulent basis, it would be difficult for either the Indonesians or the West Papuans to believe that there is any commitment to take seriously the humanity and the aspirations of the latter.

When we give priority to accompanying people and consulting them in asking what should be done for them, we shall need accurate information about their lives and the concrete impact on those lives of what is currently done to them. We need to know what happens when a flag is flown, and to enter the experience of being shot, beaten, intimidated and tortured. We need to know the human consequences of mining experienced in the surrounding towns and villages and in the valleys and fisheries downstream of the mine. We also need to know who gains financially from the counter-insurgency program and from the role of Freeport in bankrolling the activities of the security forces. We need to understand, too, the strategies devised and followed by the military in their operations in West Papua, and particularly their exploitation of social and religious divisions. Above all, we need to understand in concrete detail how these policies express themselves in the lives of West Papuans and of the Indonesian immigrants. Martinkus offers a persuasive sketch of what life is like for the people of West Papua and his essay, short though it is, suggests that they have nothing to gain from their integration into Indonesia, and that Indonesia will pay a heavy price in the corruption of its institutions for the wealth it derives from the region.

At all events, it is difficult to believe that any morally informed reflection that begins with the human reality of West Papua could support the kind of counter-insurgency policy seen recently in East Timor and now in West Papua. The involvement of the military in the political and economic life of the region and the manipulation of political and community processes for military ends have been destructive of West Papuan society. Furthermore, such power tends to corrupt both those who exercise it and those who resist it. Counter-insurgency programs are self-supporting: the violent and repressive activity of the military creates an insurgency, which is then used to justify counter-insurgency, and the subsequent priority given to security allows the military to take an even more dominant role in the political and commercial life of the region. Indonesian control of the province is secured by waging low-level warfare against people whom they insist are their citizens. For other nations to support this regime in the interests of national integrity, and even more to assist it by arms sales and assisting the military in their training is to be morally complicit in this oppression.

To insist that policy should be made after consulting people, accompanying them and after gathering and publicising information about the human reality of a situation offers a criterion for evaluating government policy. If governments were to take into account the moral dimension of policy, we might expect that they would encourage their diplomats to explore the human reality of West Papua, that they would press diplomatically and publicly the importance of consulting the people about what form of relationship to Indonesia they desire, and that they would make available to their own people the information they had gathered about Indonesian policy and its human consequences. Nothing that we know about the procedures of the Australian government in other areas where defenceless people are involved, notably in the case of the sea-borne asylum seekers, suggests that these moral imperatives are respected. It seems rather that policy decisions are taken on grounds of narrowly conceived self-interest, that diplomatic activity is little interested in the human dignity of those under threat, and that information about the abuse of human dignity is routinely withheld from the Australian public.

Of course, there is no guarantee that a policy developed from a moral perspective will be a wise policy. Policy needs also to take into account political necessities, including the limitations and likely unintended consequences of any action. It may be that in West Papua the overwhelming disparity in force between the Indonesian military and the West Papuan people, the readiness of the military to use that force and the absence of available remedies mean that it would be irresponsible now to encourage any movement towards independence. But, if so, this policy should not be defended by a higher calculus. It should be described simply as a response to unfortunate necessity, and accompanied by a full advertisement of the human reality of West Papua and by a declaration that it is intolerable. John Martinkus provides both an appropriate methodology and an appropriate rhetoric for ensuring that policy is morally grounded.

Andrew Hamilton

John Otto Ondawame

The subject of political upheavals in West Papua has been a matter of controversy among West Papuans and other interested groups for some time now. It is not surprising therefore that John Martinkus, who was nominated for a Walkley Award in 1999, should expose the systematic human rights abuses in West Papua.

John Martinkus is the author of *A Dirty Little War*, an eyewitness account of East Timor's struggle for independence. As a humanitarian and a freelance journalist, he presents a comprehensive account of the political situation in West Papua in the last four decades.

The essay illustrates how those who seek independence are killed, tortured, intimidated, executed, raped and 'disappeared' by the Indonesian military. Martinkus's work presents a significant contribution to the general debate and one which aims to reach a wider audience of academics, politicians and political commentators.

Although the essay is short and covers only essential parts of the story and most limitingly lacks a conclusion, it is worthwhile to read because it is timely and succeeds in covering even very recent events in the country. It encourages us to inform ourselves about the current political situation in West Papua, which requires an immediate solution. Most importantly, Martinkus's essay warns us to beware of repeating a similar cycle of violence to that suffered by East Timor in the post-referendum period. Although I find his ideas interesting, I do not think he presents sufficient facts to support the full implication of his arguments.

Paradise Betrayed talks about the ongoing denial of the Papuan rights to self-determination and independence, the Papuan responses and the prospects for the future. Martinkus's main focus of concern is on the relationship between the escalation of violence, the variety of human rights abuses and the increase in nationalist sentiment that is evident in many forms of political and social expression. The history of denial politics, Papuan responses under the leadership of the

OPM and its military wing, the National Liberation Army (TEPENAL), and the Presidium are surveyed, on one side, and the terror campaigns perpetrated by the Indonesian occupation forces are described at some length on the other. But these are supported by limited factual evidence. There are a few incorrect assertions and the names of the major Papuan actors in certain events are not always accurate. Despite the fact that he tends to glorify the role of leaders, whether from the OPM or the Presidium, Martinkus tries to balance his account of their respective roles with a view to the reconciliation and peace talks which must follow in the future.

Whatever its faults, the presentation of *Paradise Betrayed* is easy to understand and it is certainly transparent and educative as a piece of journalistic work. It captures the imagination of the reader and provides a clear picture of the inter-relationship between the root causes of political problems and the general response to them. The essay is informative and useful as a tool for raising public awareness, especially in Australia. John Martinkus's contribution should have a positive effect, it is liable to garner more of the international solidarity and support which today is sorely needed if there is to be any peace in West Papua. He also makes clear to us, his readers, that whatever happens to the Papuans in the short term, an independent West Papua is, sooner or later, a certainty. It is simply a matter of time.

John Otto Ondawame

Nonie Sharp

Paradise Betrayed is a disturbing eyewitness account of the plight of West Papuan people today. In taking us into their tragic world John Martinkus gives a sense of a terrible persecution that has been their lot over more than two generations. Part of the sadness of the story he tells is that despite the ever-more-embracing calamities that have befallen them, Papuans keep resisting and making desperate calls for their independence. His own dangerous mission is a personal answer to their cries of anguish.

Papuans continue to put their case in dangerous circumstances: to go off to fight in the bush and to risk torture and death to raise their flag, its main emblem the Morning Star. Why do they do this? Martinkus asks. Why keep protesting year after year, generation upon generation, seemingly offering themselves like lambs to a slaughter?

He gives a clue – but only a clue – to the answer. Martinkus senses how Papuans feel about the Morning Star flag; and he suggests that raising it takes on an almost religious quality in a quest for freedom.

A fuller answer calls for an understanding of their sense of who they are and wish to be; it relates to a Papuan identity that has been growing in substance and form over many generations. The Morning Star is the symbol of that identity throughout West Papua. Behind that lies an ancient mythical-religious narrative about the return of the god *Manseren* and the coming of *Koreri*, the kingdom of heaven on the soil of West Papua. In a wonderful epic poem of forty-two stanzas the myth of *Manseren* tells of the descent of the Morning Star to the earth and how the voice of a spirit from the land of souls told him of the promise of *Koreri*. In a line that evokes the prophet Moses on Mt Sinai the poem tells of the beginning of a long journey. 'For my eyes have beheld the Morning Star,' sings *Manseren* as he begins his earthly search for *Koreri*.

In a strange and contradictory way the Morning Star has gathered meaning throughout West Papua *even* as the situation has become radically and tragically worse. To know something of this is to begin to understand why Papuans refuse to give up.

Every people's struggle has its own unique motif, its own journey, its own victories and defeats. The life and work of Arnold Ap symbolises the Papuan journey. Through the musical group he formed at Cenderawasih University known as *Mambesak* (Shining Bird) he gave confidence and hope to thousands of his people. A unique and active bearer of Papuan culture, Arnold Ap was murdered by the Indonesian military in August 1984. He was killed because he was an upholder of *Koreri* and the spirit of Papua symbolised in the Morning Star.

The late Markus Kaisiëpo, a West Papuan leader living in exile with whom I wrote *The Morning Star in Papua Barat* in 1994, explains what Arnold Ap's life was about. 'He made *Koreri* live again.' West Papuans are *Koreri* people; it is their heritage and their hope. A man of immense stature and creativity, Ap was killed because as an intellectual imbued with the spirit of *Koreri* he gave people a belief in themselves, a feeling of hope and a desire to struggle in the face of the brutal denigration of their customs and beliefs.

Martinkus mentions a wave of 1,000 refugees arriving at the Papua New Guinea border in 1977 and the events that led to their uprooting. This tragedy produced an urgent and passionate response in me. A small book, *The Rule of the Sword: The Story of West Irian*, was the outcome. The response was electric. It surged along two quite different paths. The book broke through a silence that was not meant to be broken. The book passed from hand to hand in West Papua (Irian as it was known then). People wrote me furtive, poignant letters of thanks, even surprise, beseeching me to keep telling their story. Only later when I began to see the meaning of the Morning Star for Papuans did I realise how they must have felt when they saw the cover of my book with the Morning Star flag stretched across the map of West Papua. I recalled the stories of Australian journalist Hugh Lunn and historian Charles Rowley recounting Papuan pleas to them during that pitiful and brutal mockery of a free vote — the Act of Free Choice in 1969 where, watched by soldiers with guns, selected representatives cast their votes to stay within Indonesia. They found letters hidden in baler shells in their hotel rooms; in perilous circumstances messages were passed to them on scraps of paper. Eight years later many Papuans were fleeing after new reprisals.

At that time a quite different highly charged response came from Australian upholders of the status quo. The West Papuan cause was lost, invisible and unmentionable. To explain, publicise, support its moral rightness would only cause more suffering. Better to encourage Papuans to adjust and temporise.

(Would they tell that to an Irish rebel in Dublin or Derry? I wondered then; say that to a Palestinian on the West Bank today?)

In those days there was no transmigration of Javanese people on to Papuan lands, no pro-independence militias. And Freeport, the vast gold and copper mine, that 'emblem of globalised greed' — was only a 'promise' on a piece of paper. But there was certainly an OPM! The *Organisasi Papua Merdeka* or Free Papua Movement formed in 1965.

Today a new generation stands beside the remnants of an older one. And perhaps inspired by the Timorese many of those once prepared to put up with their subjection find they can no longer do so. I do not think that island was ever a paradise, but it is a land with its own human soul. James McAuley's poem 'New Guinea' is about the bird-shaped island, 'Land of apocalypse where the earth dances'. It was about the native-born, their path towards the unknown. The last forty years has given a new meaning to the western half — an apocalypse unforeseen by the poet. Sometimes the earth still dances but it is often dark there now. It is a land stolen, defaced, mutilated. This is the story John Martinkus tells. He ventured into a dangerous and often dark abyss to get his story. Everyone should read and reflect upon it.

Nonie Sharp

John Martinkus

When I began to research the essay *Paradise Betrayed* in February of this year I saw the piece as a chance to update the current view of the situation. The last comprehensive book published on the subject of the OPM and West Papua was Robin Osborne's *Indonesia's Secret War: The Guerrilla Struggle in Irian Jaya*, published in 1985. Of course, academic writing and news reports have continued to come out of West Papua since then but there has been nothing comprehensive that reflected the actual opinions of the people who live there and who perpetuate the struggle for independence by resisting Indonesian rule.

That was what I was trying to address and that was why I went to West Papua to carry out the research for the essay. Of course, having experienced East Timor's transition to independence through living in Dili and working as a correspondent from mid-1998 until early 2000 has shaped my view of the Indonesian military. When I approached the subject and the reality of West Papua I was coming from a perspective formed by seeing the brutality of the Indonesian military at very close range. That's something that I think some of the authors of these responses have not done – otherwise they would not be so quick to decide what level of repression it is acceptable for other people to live under.

Bruce Grant and John Ondawame both make the point that the essay has no ending or conclusion. It is a valid point but the obvious answer is that the West Papuan story has not yet run its course. I was trying to keep the essay as current and as timely as possible. If I were writing the conclusion today I would probably end on an equally inconclusive but perhaps more pessimistic note, on account of the renewed calls for Australia–Indonesia military relations to be resumed in the wake of the Bali bombings. This can only mean in the short term that the issue of self-determination for the West Papuan people will be pushed further back on the agenda of the international community. It will also mean the routine brutality of the Indonesian security forces in both West Papua and Aceh will not be criticised. It is a disturbing signal when someone of no less authority than the

United States Secretary of State, Colin Powell, can make the comment that human rights abuses by the Indonesian military will have to be overlooked in the short term because of the broader concerns of the War on Terror.

In late December 2001, Indonesian President Megawati Sukarnoputri told military leaders in Jakarta that they need not worry about violating human rights. Addressing the senior military commanders in the presence of thousands of assembled troops at a military parade, she said, 'Suddenly we are aware … of the need of a force to protect our beloved nation and motherland from breaking up.' She told them to respect the law in the course of their duty. 'With that as your guide, you can do your duty without worrying about being involved in human rights abuses,' she said. 'Have no doubts about anything you do.' Such sentiments reveal how little Indonesia has heeded international criticism about the military's abuses and how, in the wake of the Bali bombings, the international community is letting them get away with it.

Grant talks about the development of international law as a framework in which the West Papuan issue needs to be resolved and he uses the example of Milosevic to show why this is the case. I wonder why he doesn't mention the impotence of international law when it comes to the atrocities in East Timor. No Indonesian military personnel have yet been held accountable for what happened in 1999. They were all acquitted in the Jakarta trials and nothing has yet been put in place to follow that up as the United Nations pledged it would back in 1999. It would seem the signal to the Indonesian military from the international community is that they will enjoy continued impunity, not responsibility, for their actions. That is the reality that has encouraged them to act in West Papua and Aceh as they are now doing.

Grant also talks about the support the Australian media gave to East Timor and how a sustained campaign to avenge the deaths of the Balibo Five kept East Timor in the news. As someone who reported on East Timor in the media I can recall the very opposite of this being true in the mid-nineties. It was extremely difficult to publish anything dealing with East Timor's struggle for independence at that stage. Quite apart from that, it is interesting how Grant reflects that when the Santa Cruz massacre in 1991 forced a re-think on East Timor policy, 'many still preferred a form of autonomy for a period of perhaps five years'. The same thing seems to be happening now with West Papua. The people in West Papua make their rejection of autonomy quite clear through constant repetition. They don't want it. They want independence. Whether that is viable or not is irrelevant because it is what they want. By reporting the fact of that desire I am only saying what is true on the ground, but more of that later. The interesting parallel is

that, while sections of the East Timorese leadership agreed with autonomy in late
'98, as soon as the option of independence was offered by Habibie in early '99
they jumped at it. Exactly as the West Papuans would if given the chance. The
point being that autonomy will not satisfy the demands of the West Papuans and
therefore cannot be touted as a solution to the problem of West Papua. It simply
will not stop the calls for independence that are at the heart of the conflict in
West Papua.

The suggestion of UN civilian police being deployed in West Papua to create
a buffer between the Indonesian authorities and the OPM is interesting but it too
is not a solution that will be implemented in the short term. What is more likely is the other scenario Grant suggests: that the state will have to lose control,
law and order break down and human rights violations reach massive proportions before any kind of international intervention is considered. Unfortunately
I think that is the more realistic prediction given the current Australian head in
the-sand policy approach to West Papua — an approach that (as I said in the
essay) will mean that at some stage in the future the Australian military will be
called upon to perform a similar role to the one it performed in East Timor.

Andrew Hamilton makes the point that commentators 'make little of the murder, rape, assault, extortion and despoliation' that characterise current Indonesian
policy. He underlines how these issues are packaged as human rights issues, 'to
be stored at the back of the shop in case someone should ask about them'.
Reading that, I immediately think of Chris Ballard's response. He seems to be particularly affronted by the fact that I have spoken to elements of the OPM in the
mountains and calls figures I received from the OPM regarding Indonesian atrocities as 'suspiciously precise'. He then goes on to say that my assumptions about
the autonomy program — being driven as it is by those groups who stand to benefit from it — are 'naive and unfair'. I think, despite Ballard's extensive work in
West Papua, he is himself somewhat naive in his faith in the incorruptibility of
Papuans involved in the autonomy process, let alone the Indonesian bureaucracy.
He is also very innocent to imagine that in an environment where extra-judicial
executions, rape and torture have been routinely carried out by the security forces
in West Papua, that the Indonesian authorities will not use the disbursement of
autonomy funds as another coercive tool to divide Papuans and try to ensure the
loyalty of various individuals to the Indonesian state.

While I am on the subject of Ballard's letter, I cannot pass by his assertion
that it is inappropriate for foreigners to call for independence in Papua because
we are not citizens of Indonesia. To carry that statement through to its logical
conclusion is like saying that one cannot condemn the Holocaust if one was not

a citizen of the Third Reich. He goes on to say that it is appropriate for us to work for basic human rights and for the right to self-determination for West Papuans. As I see it, he is doing precisely what Hamilton put his finger on in his response. Ballard is separating the consequences of Indonesian rule from any questioning of the legitimacy of that rule and you just can't do that. Because one thing causes the other – they are intrinsically linked. Human rights do not exist in a vacuum and there is no great secret about who is abusing human rights in West Papua and why. For Christ's sake, let's speak plainly about what is going on there. The people don't want to be part of Indonesia and that is why the Indonesian state is trying to coerce them through violence into a position they do not want to accept – which is to say the permanent acknowledgement of Indonesian sovereignty.

Ballard also refers to what he calls my introduction of race into the argument for West Papuan independence. I am merely acknowledging what is blatantly true and what is evident to even the most casual observer. The Papuan people see themselves as distinct from the majority of Indonesians. Yes, I agree with him that this is partly because of the racism to which they are subject from the Indonesians, but it is also quite simply because they are a different race and that contributes to their separate identity.

It is odd that Ballard gets himself into a lather about the phrase 'the next East Timor'. An article I wrote in *The Bulletin* about the continuing Indonesian military operations in Aceh in December 2000 referred to the situation as 'the next East Timor'. It is journalistic short-hand for revealing that another conflict is taking place in which the Indonesian military is carrying out human rights abuses against a population seeking independence. And it also implies that the impact on Australia will be just as serious in terms of future military and humanitarian assistance if the Australian government doesn't recognise the inevitability of conflict in West Papua assuming Indonesia continues with its repressive policies. And what kind of journalism Mr Ballard thinks is appropriate for West Papua is neither here nor there. Does he want people to try to report conflicts without talking to the participants? What is it about conflicts in 'Indonesia' that fosters this overcompensation, this sense of the sanctity of the Indonesian state's views over those of the people it oppresses? Or is it more to do with the well-established tradition of Australian governments, Australian media and academics making excuses for the excesses of the Indonesian military? A reporter in Rwanda, for example, would not have been expected to qualify statements from a massacre survivor by speaking directly to the Hutu militia. Why is it that in Australia we have this obsession with belittling the suffering of Indonesia's victims and making excuses for the Indonesian military? The Former US Ambassador to Indonesia Robert Gelbard came out with a similar reflection when he was interviewed by SBS *Dateline* after the Bali bombing:

> Too many of the so-called experts on Indonesia, including academics in Australia and the United States and others, always say we should be very gentle with the Indonesians and not pressure them. I think a lot of the result of that has been that, too often, the Indonesians have felt a sense of what I call 'unconditional entitlement', that they can take action or not take action and get away with them. As they actually did in East Timor, for example, where they ultimately were not held tremendously accountable even when trials resulted in people being let off.

Gelbard was talking about why the Indonesian state had ignored US warnings about terrorist threats but it applies equally to the reporting of conflicts within Indonesian territories.

Which brings me to Griffin's letter. Leaving aside the rather patronising tone of his criticism, I think he provides some interesting information about Australia's position regarding West Papua in the 1950s — but his knowledge of recent events in Indonesia appears to be a bit sketchy. He questions the validity of East Timor's break away from Indonesia as a precedent for what is happening in West Papua now and he is dead wrong. I could not disagree more strongly. I think the fact that East Timor is now a successful independent nation and the fact that this has happened in our region with the support of the international community has changed the long-term outlook for West Papua immensely. The international condemnation of what the Indonesian military did in East Timor has opened the eyes of many in the United Nations and the US to the nature of the Indonesian military and the institution of the Indonesian state. To try and ignore what is probably the most significant event in our region in the last twenty-five years — the unprecedented reversal of a brutal military occupation — shows a lamentably shortsighted view of what is happening in Indonesia today.

And, yes, Professor Griffin, I do think working as a journalist in East Timor and Aceh since 1995 is preparation enough for tackling the issue of West Papua's struggle for independence, and I always tape interviews not conducted in English to review the translations later. If Griffin read the essay closely he would have seen that I make a distinction between the attitude to West Papuans of the average PNG citizen and that of the authorities in PNG and why. And if he read the notes on sources he would have seen that I acknowledge Indonesia's intention to acquire West Papua as early as 1945.

Nonie Sharp recalls how her publication of *The Rule of the Sword: The Story of West Irian* led to a highly charged reaction from Australian supporters of Indonesia's takeover of West Papua. 'The West Papuan cause was lost, invisible and unmentionable. To explain, publicise, support its moral rightness would only cause more suffering. Better to encourage Papuans to adjust and temporise.' It is a familiar sentiment and something that those who encourage Papuans to accept autonomy now, against the widespread objections of the people whom it is supposed to benefit, are replicating. There is something paternalistic in implying that the Papuan struggle for independence should not be encouraged or even recognised for what it is. And it is also morally wrong to deny one set of people freedom from foreign domination while applauding another in East Timor who have broken away from the same occupying power.

John Martinkus

Chris Ballard is a Fellow in the Research School of Pacific and Asian Studies at the Australian National University. He is one of the co-ordinators of Papuaweb, a new research website on Papua (www.papuaweb.org).

Bruce Grant is a former ambassador and government adviser, currently adjunct professor at Monash University, teaching statecraft and diplomacy. His first book, *Indonesia*, became a classic. His latest book is *A Furious Hunger: America and the 21st Century*.

James Griffin's interest in what was originally called the West New Guinea problem goes back to the 1950s. He taught at the University of Papua New Guinea for fifteen years from 1968 and 1990 and is the author of *Papua New Guinea: A Political History*.

Andrew Hamilton, S.J. teaches at the United Faculty of Theology in Melbourne and is the publisher of *Eureka Street*.

Sylvia Lawson's books include *The Archibald Paradox: A Strange Case of Authorship* (winner of the ASAL Walter McCrae Russell Award) and, most recently, *How Simone de Beauvoir Died in Australia*.

Amanda Lohrey began her working life as a lecturer in political science at the University of Tasmania. She has since published a number of articles and essays on Australian political life as well as two political novels, *The Morality of Gentlemen* and *The Reading Group*. Her most recent novel is the award-winning *Camille's Bread*.

John Martinkus is an Australian investigative reporter on the Asia region. In 1999 he was nominated for a Walkley Award for his coverage of the violence in East Timor. His book *A Dirty Little War*, an eyewitness account of East Timor's struggle for independence, was short-listed for the NSW Premier's Literary Awards in 2002.

John Ondawame is an International Spokesperson of the OPM, a member of the Papuan Presidium and the Coordinator of West Papua Project, Centre for Peace and Conflict Studies, University of Sydney.

Nonie Sharp is the author of *The Rule of the Sword: The Story of West Irian* and *The Morning Star in Papua Barat* (in association with Markus Wonggor Kasiëpo). Her latest book is *Saltwater People: The Waves of Memory*.

QUARTERLY ESSAY

SUBSCRIPTIONS Receive a discount and never miss an issue. Mailed direct to your door. 1 year subscription (4 issues): $46.95 a year within Australia incl. GST (Institutional subs. $52.95). Outside Australia $74.95. All prices include postage and handling.

BACK ISSUES Please add $2.50 postage and handling to your order (or $8.00 for overseas orders).

- ☑ **Issue 1** ($9.95) Robert Manne's *In Denial: The Stolen Generations and the Right*
- ☑ **Issue 2** ($9.95) John Birmingham's *Appeasing Jakarta: Australia's Complicity in the East Timor Tragedy*
- ☑ **Issue 3** ($9.95) Guy Rundle's *The Opportunist: John Howard and the Triumph of Reaction*
- ☑ **Issue 4** ($9.95) Don Watson's *Rabbit Syndrome: Australia and America*
- ☑ **Issue 5** ($11.95) Mungo MacCallum's *Girt by Sea: Australia, the Refugees and the Politics of Fear*
- ☐ **Issue 6** ($11.95) John Button's *Beyond Belief: What Future for Labor?*
- ☐ **Issue 7** ($11.95) John Martinkus's *Paradise Betrayed: West Papua's Struggle for Independence*

PAYMENT DETAILS I enclose a cheque/money order made out to Schwartz Publishing Pty Ltd. Please debit my credit card (Mastercard, Visa Card or Bankcard accepted).

Card No. ☐☐☐☐☐☐☐☐☐☐☐☐☐☐☐☐

Expiry date ___ / ___ Amount $ __________

Cardholder's name __________

Signature __________

Name __________

Address __________

____

Email __________

POST OR FAX TO:
Black Inc.
Level 5, 289 Flinders Lane, Melbourne,
Victoria 3000 Australia
Telephone: 61 3 9654 2000
Facsimile: 61 3 9654 2290
Email: quarterlyessay@blackincbooks.com

Subscribe online at www.blackincbooks.com